Theoretical and Mathematical Physics

The series founded in 1975 and formerly (until 2005) entitled *Texts and Monographs in Physics* (TMP) publishes high-level monographs in theoretical and mathematical physics. The change of title to *Theoretical and Mathematical Physics* (TMP) signals that the series is a suitable publication platform for both the mathematical and the theoretical physicist. The wider scope of the series is reflected by the composition of the editorial board, comprising both physicists and mathematicians.

The books, written in a didactic style and containing a certain amount of elementary background material, bridge the gap between advanced textbooks and research monographs. They can thus serve as basis for advanced studies, not only for lectures and seminars at graduate level, but also for scientists entering a field of research.

T0155794

John von Neumann

Claude Shannon

Erwin Schrödinger

Dénes Petz

Quantum Information Theory and Quantum Statistics

With 10 Figures

 Springer

Prof. Dénes Petz
Alfréd Rényi Institute of Mathematics
POB 127, H-1364 Budapest, Hungary
petz@math.bme.hu

D. Petz, *Quantum Information Theory and Quantum Statistics*, Theoretical and Mathematical Physics (Springer, Berlin Heidelberg 2008) DOI 10.1007/978-3-540-74636-2

ISBN 978-3-642-09409-5   e-ISBN 978-3-540-74636-2

Theoretical and Mathematical Physics ISSN 1864-5879

Cover design: eStudio Calamar, Girona/Spain

Printed on acid-free paper

9 8 7 6 5 4 3 2 1

springer.com

Preface

Quantum mechanics was one of the very important new theories of the 20th century. John von Neumann worked in Göttingen in the 1920s when Werner Heisenberg gave the first lectures on the subject. Quantum mechanics motivated the creation of new areas in mathematics; the theory of linear operators on Hilbert spaces was certainly such an area. John von Neumann made an effort toward the mathematical foundation, and his book "*The mathematical foundation of quantum mechanics*" is still rather interesting to study. The book is a precise and self-contained description of the theory, some notations have been changed in the mean time in the literature.

Although quantum mechanics is mathematically a perfect theory, it is full of interesting methods and techniques; the interpretation is problematic for many people. An example of the strange attitudes is the following: "*Quantum mechanics is not a theory about reality, it is a prescription for making the best possible prediction about the future if we have certain information about the past*" (G. 't' Hooft, 1988). The interpretations of quantum theory are not considered in this book. The background of the problems might be the probabilistic feature of the theory. On one hand, the result of a measurement is random with a well-defined distribution; on the other hand, the random quantities do not have joint distribution in many cases. The latter feature justifies the so-called quantum probability theory.

Abstract information theory was proposed by electric engineer Claude Shannon in the 1940s. It became clear that coding is very important to make the information transfer efficient. Although quantum mechanics was already established, the information considered was classical; roughly speaking, this means the transfer of 0–1 sequences. Quantum information theory was born much later in the 1990s. In 1993 C. H. Bennett, G. Brassard, C. Crepeau, R. Jozsa, A. Peres and W. Wootters published the paper *Teleporting an unknown quantum state via dual classical and EPR channels*, which describes a state teleportation protocol. The protocol is not complicated; it is somewhat surprising that it was not discovered much earlier. The reason can be that the interest in quantum computation motivated the study of the transmission of quantum states. Many things in quantum information theory is related to quantum computation and to its algorithms. Measurements on a quantum system provide classical information, and due to the randomness classical statistics can be used to estimate the true state. In some examples, quantum information can appear, the state of a subsystem can be so.

The material of this book was lectured at the Budapest University of Technology and Economics and at the Central European University mostly for physics and mathematics majors, and for newcomers in the area. The book addresses graduate students in mathematics, physics, theoretical and mathematical physicists with some interest in the rigorous approach. The book does not cover several important results in quantum information theory and quantum statistics. The emphasis is put on the real introductory explanation for certain important concepts. Numerous examples and exercises are also used to achieve this goal. The presentation is mathematically completely rigorous but friendly whenever it is possible. Since the subject is based on non-trivial applications of matrices, the appendix summarizes the relevant part of linear analysis. Standard undergraduate courses of quantum mechanics, probability theory, linear algebra and functional analysis are assumed. Although the emphasis is on quantum information theory, many things from classical information theory are explained as well. Some knowledge about classical information theory is convenient, but not necessary.

I thank my students and colleagues, especially Tsuyoshi Ando, Thomas Baier, Imre Csiszár, Katalin Hangos, Fumio Hiai, Gábor Kiss, Milán Mosonyi and József Pitrik, for helping me to improve the manuscript.

Dénes Petz

Contents

Chapter 1
Introduction

Given a set $\mathscr{X}$ of outcomes of an experiment, information gives one of the possible alternatives. The value (or measure) of the information is proportional to the size of $\mathscr{X}$. The idea of measuring information regardless of its content dates back to **R. V. L. Hartley** (1928). He recognized the logarithmic nature of the natural measure of information content: When the cardinality of $\mathscr{X}$ is n, the amount of information assigned to an element is $\log n$. (The base of the logarithm yields a constant factor.)

When a probability mass function $p(x)$ is given on $\mathscr{X}$, then the situation is slightly more complicated. **Claude Shannon** proposed the formula

$$H(p) = - \sum_{x \in \mathscr{X}} p(x) \log_2 p(x),$$

so he used logarithm to the base 2 and he called this quantity "entropy". Assume that $\#(\mathscr{X}) = 8$ and the probability distribution is uniform. Then

$$H = - \sum \frac{1}{8} \log_2 \frac{1}{8} = 3,$$

in accordance with the fact that we need 3-bit strings to label the elements of $\mathscr{X}$.

Suppose eight horses take part in a race and the probabilities of winning are

$$\left(\frac{1}{2}, \frac{1}{4}, \frac{1}{8}, \frac{1}{16}, \frac{1}{64}, \frac{1}{64}, \frac{1}{64}, \frac{1}{64} \right). \tag{1.1}$$

If we want to inform somebody about the winner, then a possibility is to send the index of the winning horse. This protocol requires 3 bits independently of the actual winner. However, it is more appropriate to send a shorter message for a more probable horse and to send a longer for the less probable one. If we use the strings

$$0, 10, 110, 1110, 111100, 111101, 111110, 111111, \tag{1.2}$$

then the average message length is 2 bits. This coincides with the Shannon entropy of the probability distribution (1.1) and smaller than the uniform code length. From

D. Petz, *Introduction*. In: D. Petz, Quantum Information Theory and Quantum Statistics, Theoretical and Mathematical Physics, pp. 1–2 (2008)
DOI 10.1007/978-3-540-74636-2_1
© Springer-Verlag Berlin Heidelberg 2008

the example, one observes that coding can make the information transfer more efficient. The Shannon entropy is a theoretical lower bound for the average code length. It is worthwhile to note that the coding (1.2) has a very special property. If the race is repeated and the two winners are messaged by a sequence

$$11110010,$$

then the first and the second winners can be recovered uniquely:

$$111100 + 10.$$

Physics of the media that is used to store or transfer information determines how to manipulate that information. Classical media, for example magnetic domains or a piece of paper, determine classical logic as the means to manipulate that information. In classical logic, things are true or false; for example, a magnetic domain on the drive either is aligned with the direction of the head or is not. Any memory location can be read without destroying that memory location. A physical system obeying the laws of quantum mechanics is rather different. Any measurement performed on a quantum system destroys most of the information contained in that system. The discarded information is unrecoverable. The outcome of the measurement is stochastic, in general probabilistic predictions can be made only. A quantum system basically carries quantum information, but the system can be used to store or transfer classical information as well.

The word *teleportation* is from science fiction and it means that an object disintegrates in one place while a perfect replica appears somewhere else. The quantum teleportation protocol discovered in 1993 by Richard Jozsa, William K. Wootters, Charles H. Bennett, Gilles Brassard, Claude Crépeau and Asher Peres is based on 3 quantum bits. Alice has access to the quantum bits X and A, Bob has the quantum bit B. The bits A and B are in a special quantum relation, they are **entangled**. This means that the bit B senses if something happens with the bit A. The state of the bit X is not known, and the goal is to transfer its state to Bob. Alice performs a particular measurement on the quantum bits X and A. Her measurement has 4 different outcomes and she informs Bob about the outcome. Bob has a prescription: each of the four outcomes corresponds to a dynamical change of the state of the quantum bit B. He performs the change suggested by Alice's information. After that the state of the quantum bit B will be exactly the same as the state of the quantum bit X before Alice' measurement. The teleportation protocol is not in contradiction to the uncertainty principle. Alice does not know the initial state of X and Bob does not know the final state of B. Nevertheless, the two states are exactly the same.

The teleportation protocol is based on entanglement. The state of a quantum bit is described by a 2×2 positive semidefinite matrix which has complex entries and trace 1. Such a matrix is determined by 3 real numbers. Two quantum bits form a 4-level-quantum system, and the description of a state requires 15 real numbers. One can argue that $15 - 2 \times 3 = 9$ numbers are needed to describe the relation of the two qubits. The relation can be very complex, and entanglement is an interesting and important example.

Chapter 2
Prerequisites from Quantum Mechanics

The starting point of the quantum mechanical formalism is the **Hilbert space**. The Hilbert space is a mathematical concept, it is a space in the sense that it is a complex vector space which is endowed by an **inner** or **scalar product** $\langle \cdot, \cdot \rangle$. The linear space $\mathbb{C}^n$ of all n-tuples of complex numbers becomes a Hilbert space with the inner product

$$\langle x, y \rangle = \sum_{i=1}^{n} \bar{x}_i y_i = [\bar{x}_1, \bar{x}_2, \ldots \bar{x}_n] \begin{bmatrix} y_1 \\ y_2 \\ \cdot \\ \cdot \\ y_n \end{bmatrix},$$

where $\bar{z}$ denotes the complex conjugate of the complex number $z \in \mathbb{C}$. Another example is the space of square integrable complex-valued functions on the real Euclidean space $\mathbb{R}^n$. If f and g are such functions then

$$\langle f, g \rangle = \int_{\mathbb{R}^n} \overline{f(x)} g(x) \, dx$$

gives the inner product. The latter space is denoted by $L^2(\mathbb{R}^n)$ and it is infinite dimensional contrary to the n-dimensional space $\mathbb{C}^n$. We are mostly satisfied with finite dimensional spaces. The inner product of the vectors $|x\rangle$ and $|y\rangle$ will be often denoted as $\langle x|y\rangle$; this notation, sometimes called "bra" and "ket," is popular in physics. On the other hand, $|x\rangle \langle y|$ is a linear operator which acts on the ket vector $|z\rangle$ as

$$\left(|x\rangle \langle y| \right) |z\rangle = |x\rangle \langle y|z\rangle \equiv \langle y|z\rangle |x\rangle.$$

Therefore,

$$|x\rangle \langle y| = \begin{bmatrix} x_1 \\ x_2 \\ \cdot \\ \cdot \\ x_n \end{bmatrix} [\bar{y}_1, \bar{y}_2, \ldots \bar{y}_n]$$

is conjugate linear in $|y\rangle$, while $\langle x|y\rangle$ is linear.

D. Petz, *Prerequisites from Quantum Mechanics*. In: D. Petz, Quantum Information Theory and Quantum Statistics,
Theoretical and Mathematical Physics, pp. 3–24 (2008)
DOI 10.1007/978-3-540-74636-2_2   © Springer-Verlag Berlin Heidelberg 2008

In this chapter I explain shortly the fundamental postulates of quantum mechanics about quantum states, observables, measurement, composite systems and time development.

2.1 Postulates of Quantum Mechanics

The basic postulate of quantum mechanics is about the Hilbert space formalism.

(A0)   To each quantum mechanical system a complex Hilbert space $\mathcal{H}$ is associated.

The (pure) physical states of the system correspond to unit vectors of the Hilbert space. This correspondence is not 1–1. When f_1 and f_2 are unit vectors, then the corresponding states are identical if $f_1 = zf_2$ for a complex number z of modulus 1. Such z is often called **phase**. The **pure physical state** of the system determines a corresponding state vector up to a phase.

Example 2.1. The two-dimensional Hilbert space $\mathbb{C}^2$ is used to describe a 2-level quantum system called **qubit**. The canonical basis vectors $(1,0)$ and $(0,1)$ are usually denoted by $|\uparrow\rangle$ and $|\downarrow\rangle$, respectively. (An alternative notation is $|1\rangle$ for $(0,1)$ and $|0\rangle$ for $(1,0)$.) Since the polarization of a photon is an important example of a qubit, the state $|\uparrow\rangle$ may have the interpretation that the "polarization is vertical" and $|\downarrow\rangle$ means that the "polarization is horizontal".

To specify a state of a qubit we need to give a real number x_1 and a complex number z such that $x_1^2 + |z|^2 = 1$. Then the state vector is

$$x_1\,|\uparrow\rangle + z\,|\downarrow\rangle.$$

(Indeed, multiplying a unit vector $z_1\,|\uparrow\rangle + z_2\,|\downarrow\rangle$ by an appropriate phase, we can make the coefficient of $|\uparrow\rangle$ real and the corresponding state remains the same.)

Splitting z into real and imaginary parts as $z = x_2 + ix_3$, we have the constraint $x_1^2 + x_2^2 + x_3^2 = 1$ for the parameters $(x_1, x_2, x_3) \in \mathbb{R}^3$.

Therefore, the space of all pure states of a qubit is conveniently visualized as the sphere in the three-dimensional Euclidean space; it is called the **Bloch sphere**.   □

Traditional quantum mechanics distinguishes between pure states and **mixed states**. Mixed states are described by **density matrices**. A density matrix or statistical operator is a positive operator of trace 1 on the Hilbert space. This means that the space has a basis consisting of weigenvectors of the statistical operator and the sum of eigenvalues is 1. (In the finite dimensional case the first condition is automatically fulfilled.) The pure states represented by unit vectors of the Hilbert space are among the density matrices under an appropriate identification. If $x = |x\rangle$ is a unit vector, then $|x\rangle\langle x|$ is a density matrix. Geometrically $|x\rangle\langle x|$ is the orthogonal projection onto the linear subspace generated by x. Note that $|x\rangle\langle x| = |y\rangle\langle y|$ if the vectors x and y differ in a phase.

(A1) The physical states of a quantum mechanical system are described by statistical operators acting on the Hilbert space.

Example 2.2. A state of the spin (of $1/2$) can be represented by the 2×2 matrix

$$\frac{1}{2} \begin{bmatrix} 1+x_3 & x_1-ix_2 \\ x_1+ix_2 & 1-x_3 \end{bmatrix}. \tag{2.1}$$

This is a density matrix if and only if $x_1^2 + x_2^2 + x_3^2 \le 1$ (Fig. 2.1). $\qquad\square$

The second axiom is about observables.

(A2) The observables of a quantum mechanical system are described by self-adjoint operators acting on the Hilbert space.

A **self-adjoint operator** A on a Hilbert space $\mathscr{H}$ is a linear operator $\mathscr{H} \to \mathscr{H}$ which satisfies

$$\langle Ax, y \rangle = \langle x, Ay \rangle$$

for $x, y \in \mathscr{H}$. Self-adjoint operators on a finite dimensional Hilbert space $\mathbb{C}^n$ are $n \times n$ self-adjoint matrices. A self-adjoint matrix admits a **spectral decomposition**

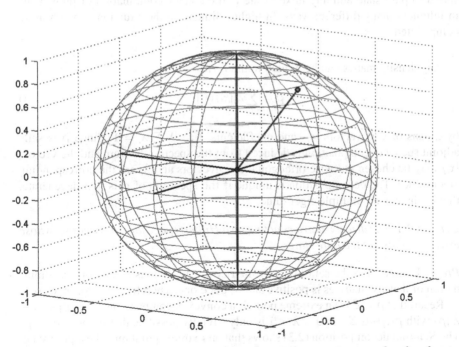

Fig. 2.1 A 2×2 density matrix has the form $\frac{1}{2}(I + x_1\sigma_1 + x_2\sigma_2 + x_3\sigma_3)$, where $x_1^2 + x_2^2 + x_3^2 \le 1$. The length of the vectors (x_1, x_2, x_3) is at most 1 and they form the unit ball, called Bloch ball, in the three-dimensional Euclidean space. The pure states are on the surface

$A = \sum_i \lambda_i E_i$, where λ_i are the different eigenvalues of A, and E_i is the orthogonal projection onto the subspace spanned by the eigenvectors corresponding to the eigenvalue λ_i. Multiplicity of λ_i is exactly the rank of E_i.

Example 2.3. In case of a quantum spin (of $1/2$) the matrices

$$\sigma_1 = \begin{bmatrix} 0 & 1 \\ 1 & 0 \end{bmatrix}, \qquad \sigma_2 = \begin{bmatrix} 0 & -i \\ i & 0 \end{bmatrix}, \qquad \sigma_3 = \begin{bmatrix} 1 & 0 \\ 0 & -1 \end{bmatrix}$$

are used to describe the spin of direction x, y, z (with respect to a coordinate system). They are called **Pauli matrices**. Any 2×2 self-adjoint matrix is of the form

$$A_{(x_0,x)} := x_0\sigma_0 + x_1\sigma_1 + x_2\sigma_2 + x_3\sigma_3$$

if σ_0 stands for the unit matrix I. We can also use the shorthand notation $x_0\sigma_0 + x \cdot \sigma$. The density matrix (2.1) can be written as

$$\tfrac{1}{2}(\sigma_0 + x \cdot \sigma), \tag{2.2}$$

where $\|x\| \leq 1$. x is called **Bloch vector** and these vectors form the **Bloch ball**.

Formula (2.2) makes an affine correspondence between 2×2 density matrices and the unit ball in the Euclidean 3-space. The extreme points of the ball correspond to pure state and any mixed state is the convex combination of pure states in infinitely many different ways. In higher dimension the situation is much more complicated.   □

Any density matrix can be written in the form

$$\rho = \sum_i \lambda_i |x_i\rangle\langle x_i| \tag{2.3}$$

by means of unit vectors $|x_i\rangle$ and coefficients $\lambda_i \geq 0$, $\sum_i \lambda_i = 1$. Since ρ is self-adjoint such a decomposition is deduced from the spectral theorem and the vectors $|x_i\rangle$ may be chosen pairwise orthogonal eigenvectors and λ_i are the corresponding eigenvalues. The decomposition is unique if the spectrum of ρ is non-degenerate, that is, there is no multiple eigenvalue.

Lemma 2.1. *The density matrices acting on a Hilbert space form a convex set whose extreme points are the pure states.*

Proof. Denote by Σ the set of density matrices. It is obvious that a convex combination of density matrices is positive and of trace one. Therefore Σ is a convex set.

Recall that $\rho \in \Sigma$ is an extreme point if a convex decomposition $\rho = \lambda\rho_1 + (1 - \lambda)\rho_2$ with $\rho_1, \rho_2 \in \Sigma$ and $0 < \lambda < 1$ is only trivially possible, that is, $\rho_1 = \rho_2 = \rho$. The Schmidt decomposition (2.3) shows that an extreme point must be a pure state.

Let p be a pure state, $p = p^2$. We have to show that it is really an extreme point. Assume that $p = \lambda\rho_1 + (1 - \lambda)\rho_2$. Then

$$p = \lambda p\rho_1 p + (1 - \lambda)p\rho_2 p$$

and $\operatorname{Tr} p\rho_i p = 1$ must hold. Remember that $\operatorname{Tr} p\rho_i p = \langle p, \rho_i \rangle$, while $\langle p, p \rangle = 1$ and $\langle \rho_i, \rho_i \rangle \leq 1$. In the Schwarz inequality

$$|\langle e, f \rangle|^2 \leq \langle e, e \rangle \langle f, f \rangle$$

the equality holds if and only if $f = ce$ for some complex number c. Therefore, $\rho_i = c_i p$ must hold. Taking the trace, we get $c_i = 1$ and $\rho_1 = \rho_2 = p$. □

The next result, obtained by Schrödinger [105], gives relation between different decompositions of density matrices.

Lemma 2.2. *Let*

$$\rho = \sum_{i=1}^{k} |x_i\rangle\langle x_i| = \sum_{j=1}^{k} |y_j\rangle\langle y_j|$$

be decompositions of a density matrix. Then there exists a unitary matrix $(U_{ij})_{i,j=1}^{k}$ *such that*

$$\sum_{j=1}^{k} U_{ij}|x_j\rangle = |y_i\rangle. \tag{2.4}$$

Let $\sum_{i=1}^{n} \lambda_i |z_i\rangle\langle z_i|$ be the Schmidt decomposition of ρ, that is, $\lambda_i > 0$ and $|z_i\rangle$ are pairwise orthogonal unit vectors $(1 \leq i \leq n)$. The integer n is the rank of ρ, therefore $n \leq k$. Set $|z_i\rangle := 0$ and $\lambda_i := 0$ for $n < i \leq k$. It is enough to construct a unitary transforming the vectors $\sqrt{\lambda_i}|z_i\rangle$ to the vectors $|y_i\rangle$. Indeed, if two arbitrary decompositions are given and both of them connected to an orthogonal decomposition by a unitary, then one can form a new unitary from the two which will connect the two decompositions.

The vectors $|y_i\rangle$ are in the linear span of $\{|z_i\rangle : 1 \leq i \leq n\}$; therefore

$$|y_i\rangle = \sum_{j=1}^{n} \langle z_j | y_i \rangle |z_j\rangle$$

is the orthogonal expansion. We can define a matrix (U_{ij}) by the formula

$$U_{ij} = \frac{\langle z_j, |y_i \rangle}{\sqrt{\lambda_j}} \qquad (1 \leq i \leq k, 1 \leq j \leq n).$$

We can easily compute that

$$\sum_{i=1}^{k} U_{it} U_{iu}^{*} = \sum_{i=1}^{k} \frac{\langle z_t, |y_i \rangle}{\sqrt{\lambda_t}} \frac{\langle y_i, |z_u \rangle}{\sqrt{\lambda_u}}$$

$$= \frac{\langle z_t | \rho | z_u \rangle}{\sqrt{\lambda_u \lambda_t}} = \delta_{t,u},$$

and this relation shows that the n column vectors of the matrix (U_{ij}) are orthonormal. If $n < k$, then we can append further columns to get a $k \times k$ unitary. □

Quantum mechanics is not deterministic. If we prepare two identical systems in the same state, and we measure the same observable on each, then the result of the **measurement** may not be the same. This indeterminism or stochastic feature is fundamental.

(A3) Let $\mathscr{X}$ be a finite set and for $x \in \mathscr{X}$ an operator $V_x \in B(\mathscr{H})$ be given such that $\sum_x V_x^* V_x = I$. Such an indexed family of operators is a model of a measurement with values in $\mathscr{X}$. If the measurement is performed in a state ρ, then the outcome $x \in \mathscr{X}$ appears with probability $\mathrm{Tr}\, V_x \rho V_x^*$ and after the measurement the state of the system is

$$\frac{V_x \rho V_x^*}{\mathrm{Tr}\, V_x \rho V_x^*}.$$

A particular case is the measurement of an observable described by a self-adjoint operator A with spectral decomposition $\sum_i \lambda_i E_i$. In this case $\mathscr{X} = \{\lambda_i\}$ is the set of eigenvalues and $V_i = E_i$. One can compute easily that the expectation of the random outcome is $\mathrm{Tr}\, \rho A$. The functional $A \mapsto \mathrm{Tr}\, \rho A$ is linear and has two important properties: 1) If $A \geq 0$, then $\mathrm{Tr}\, \rho A \geq 0$, 2) $\mathrm{Tr}\, \rho I = 1$. These properties allow to see quantum states in a different way. If $\varphi : B(\mathscr{H}) \to \mathbb{C}$ is a linear functional such that

$$\varphi(A) \geq 0 \quad \text{if} \quad A \geq 0 \quad \text{and} \quad \varphi(I) = 1, \tag{2.5}$$

then there exists a density matrix ρ_φ such that

$$\varphi(A) = \mathrm{Tr}\, \rho_\varphi A. \tag{2.6}$$

The functional φ associates the expectation value to the observables A.

The density matrices ρ_1 and ρ_2 are called **orthogonal** if any eigenvector of ρ_1 is orthogonal to any eigenvector of ρ_2.

Example 2.4. Let ρ_1 and ρ_2 be density matrices. They can be **distinguished with certainty** if there exists a measurement which takes the value 1 with probability 1 when the system is in the state ρ_1 and with probability 0 when the system is in the state ρ_2.

Assume that ρ_1 and ρ_2 are orthogonal and let P be the orthogonal projection onto the subspace spanned by the non-zero eigenvectors of ρ_1. Then $V_1 := P$ and $V_2 := I - P$ is a measurement and $\mathrm{Tr}\, V_1 \rho_1 V_1^* = 1$ and $\mathrm{Tr}\, V_1 \rho_2 V_1^* = 0$.

Conversely, assume that a measurement (V_i) exists such that $\mathrm{Tr}\, V_1 \rho_1 V_1^* = 1$ and $\mathrm{Tr}\, V_1 \rho_2 V_1^* = 0$. The first condition implies that $V_1^* V_1 \geq P$, where P us the support projection of ρ_1, defined above. The second condition tells is that $V_1^* V_1$ is orthogonal to the support of ρ_2. Therefore, $\rho_1 \perp \rho_2$. □

Let $e_1, e_2, \ldots, e_n$ be an orthonormal basis in a Hilbert space $\mathscr{H}$. The unit vector $\xi \in \mathscr{H}$ is **complementary** to the given basis if

$$|\langle e_i, \xi \rangle| = \frac{1}{\sqrt{n}} \qquad (1 \le i \le n). \tag{2.7}$$

The basis vectors correspond to a measurement, $|e_1\rangle\langle e_1|, \ldots, |e_n\rangle\langle e_n|$ are positive operators and their sum is I. If the pure state $|\xi\rangle\langle\xi|$ is the actual state of the quantum system, then complementarity means that all outputs of the measurement appear with the same probability.

Two orthonormal bases are called **complementary** if all vectors in the first basis are complementary to the other basis.

Example 2.5. First we can note that (2.7) is equivalent to the relation

$$\text{Tr} \, |e_i\rangle\langle e_i| \, |\xi\rangle\langle\xi| = \frac{1}{n} \tag{2.8}$$

which is about the trace of the product of two projections.

The eigenprojections of the Pauli matrix σ_i are $(I \pm \sigma_i)/2$. We have

$$\text{Tr} \left(\frac{I \pm \sigma_i}{2} \frac{I \pm \sigma_j}{2} \right) = \frac{1}{2}$$

for $1 \le i \ne j \le 3$. This shows that the eigenbasis of σ_i is complementary to the eigenbasis of σ_j if i and j are different. $\qquad\qquad \square$

According to axiom (A1), a Hilbert space is associated to any quantum mechanical system. Assume that a **composite system** consists of the subsystems (1) and (2), they are described by the Hilbert spaces $\mathcal{H}_1$ and $\mathcal{H}_2$. (Each subsystem could be a particle or a spin, for example.) Then we have the following.

(A4)   The composite system is described by the tensor product Hilbert space $\mathcal{H}_1 \otimes \mathcal{H}_2$.

When $\{e_j : j \in J\}$ is a basis of $\mathcal{H}_1$ and $\{f_i : i \in I\}$ is a basis of $\mathcal{H}_2$, then $\{e_j \otimes f_i : j \in J, i \in I\}$ is a basis of $\mathcal{H}_1 \otimes \mathcal{H}_2$. Therefore, the dimension of $\mathcal{H}_1 \otimes \mathcal{H}_2$ is dim $\mathcal{H}_1 \times$ dim $\mathcal{H}_2$. If $A_i \in B(\mathcal{H}_i)$ $(i = 1, 2)$, then the action of the tensor product operator $A_1 \otimes A_2$ is determined by

$$(A_1 \otimes A_2)(\eta_1 \otimes \eta_2) = A_1 \eta_1 \otimes A_2 \eta_2$$

since the vectors $\eta_1 \otimes \eta_2$ span $\mathcal{H}_1 \otimes \mathcal{H}_2$.

When $A = A^*$ is an observable of the first system, then its expectation value in the vector state $\Psi \in \mathcal{H}_1 \otimes \mathcal{H}_2$ is

$$\langle \Psi, (A \otimes I_2)\Psi \rangle,$$

where I_2 is the identity operator on $\mathcal{H}_2$.

Example 2.6. The Hilbert space of a composite system of two spins (of $1/2$) is $\mathbb{C}^2 \otimes \mathbb{C}^2$. In this space, the vectors

$$e_1 := |\uparrow\rangle \otimes |\uparrow\rangle, \quad e_2 := |\uparrow\rangle \otimes |\downarrow\rangle, \quad e_3 := |\downarrow\rangle \otimes |\uparrow\rangle, \quad e_4 := |\downarrow\rangle \otimes |\downarrow\rangle$$

form a basis. The vector state

$$\Phi = \frac{1}{\sqrt{2}}(|\uparrow\rangle \otimes |\downarrow\rangle - |\downarrow\rangle \otimes |\uparrow\rangle) \tag{2.9}$$

has a surprising property. Consider the observable

$$A := \sum_{i=1}^{4} i|e_i\rangle\langle e_i|,$$

which has eigenvalues 1, 2, 3 and 4 and the corresponding eigenvectors are just the basis vectors. Measurement of this observable yields the values 1, 2, 3 and 4 with probabilities 0, 1/2, 1/2 and 0, respectively. The 0 probability occurs when both spins are up or both are down. Therefore in the vector state Φ the spins are anti-correlated. □

We can consider now the composite system $\mathcal{H}_1 \otimes \mathcal{H}_2$ in a state $\Phi \in \mathcal{H}_1 \otimes \mathcal{H}_2$. Let $A \in B(\mathcal{H}_1)$ be an observable which is localized at the first subsystem. If we want to consider A as an observable of the total system, we have to define an extension to the space $\mathcal{H}_1 \otimes \mathcal{H}_2$. The tensor product operator $A \otimes I$ will do, I is the identity operator of $\mathcal{H}_2$.

Lemma 2.3. *Assume that $\mathcal{H}_1$ and $\mathcal{H}_2$ are finite dimensional Hilbert spaces. Let $\{e_j : j \in J\}$ be a basis of $\mathcal{H}_1$ and $\{f_i : i \in I\}$ be a basis of $\mathcal{H}_2$. Assume that*

$$\Phi = \sum_{i,j} w_{ij} e_j \otimes f_i$$

is the expansion of a unit vector $\Phi \in \mathcal{H}_1 \otimes \mathcal{H}_2$. Set W for the matrix which is determined by the entries w_{kl}. Then W^*W is a density matrix and

$$\langle \Phi, (A \otimes I)\Phi \rangle = \mathrm{Tr}\, AW^*W.$$

Proof. Let E_{kl} be an operator on $\mathcal{H}_1$ which is determined by the relations $E_{kl}e_j = \delta_{lj}e_k$ $(k,l \in I)$. As a matrix, E_{kl} is called "matrix unit"; it is a matrix such that (k,l) entry is 1, all others are 0. Then

$$\langle \Phi, (E_{kl} \otimes I)\Phi \rangle = \left\langle \sum_{i,j} w_{ij} e_j \otimes f_i, (E_{kl} \otimes I) \sum_{t,u} w_{tu} e_u \otimes f_t \right\rangle =$$

$$= \sum_{i,j} \sum_{t,u} \overline{w}_{ij} w_{tu} \langle e_j, E_{kl} e_u \rangle \langle f_i, f_t \rangle =$$

$$= \sum_{i,j} \sum_{t,u} \overline{w}_{ij} w_{tu} \, \delta_{lu} \delta_{jk} \delta_{it} = \sum_i \overline{w}_{ik} w_{il}.$$

Then we can arrive at the (k, l) entry of W^*W. Our computation may be summarized as

$$\langle \Phi, (E_{kl} \otimes I)\Phi \rangle = \operatorname{Tr} E_{kl}(W^*W) \qquad (k, l \in I).$$

Since any linear operator $A \in B(\mathcal{H}_1)$ is of the form $A = \sum_{k,l} a_{kl} E_{kl}$ $(a_{kl} \in \mathbb{C})$, taking linear combinations of the previous equations, we have

$$\langle \Phi, (A \otimes I)\Phi \rangle = \operatorname{Tr} A(W^*W).$$

W^*W is obviously positive and

$$\operatorname{Tr} W^*W = \sum_{i,j} |w_{ij}|^2 = \|\Phi\|^2 = 1.$$

Therefore it is a density matrix.   □

This lemma shows a natural way from state vectors to density matrices. Given a density matrix ρ on $\mathcal{H}_1 \otimes \mathcal{H}_2$, there are density matrices $\rho_i \in B(\mathcal{H}_i)$ such that

$$\operatorname{Tr}(A \otimes I)\rho = \operatorname{Tr} A \rho_1 \qquad (A \in B(\mathcal{H}_1)) \tag{2.10}$$

and

$$\operatorname{Tr}(I \otimes B)\rho = \operatorname{Tr} B \rho_2 \qquad (B \in B(\mathcal{H}_2)). \tag{2.11}$$

ρ_1 and ρ_2 are called **reduced density matrices**. (They are the quantum analogue of marginal distributions.)

The proof of Lemma 2.3 contains the reduced density of $|\Phi\rangle\langle\Phi|$ on the first system; it is W^*W. One computes similarly the reduced density on the second subsystem; it is $(WW^*)^T$, where X^T denotes the transpose of the matrix X. Since W^*W and $(WW^*)^T$ have the same non-zero eigenvalues, the two subsystems are very strongly connected if the total system is in a pure state.

Let $\mathcal{H}_1$ and $\mathcal{H}_2$ be Hilbert spaces and let $\dim \mathcal{H}_1 = m$ and $\dim \mathcal{H}_2 = n$. It is well known that the matrix of a linear operator on $\mathcal{H}_1 \otimes \mathcal{H}_2$ has a block-matrix form,

$$U = (U_{ij})_{i,j=1}^m = \sum_{i,j=1}^m E_{ij} \otimes U_{ij},$$

relative to the lexicographically ordered product basis, where U_{ij} are $n \times n$ matrices. For example,

$$A \otimes I = (X_{ij})_{i,j=1}^m, \quad \text{where} \quad X_{ij} = A_{ij} I_n$$

and

$$I \otimes B = (X_{ij})_{i,j=1}^m, \quad \text{where} \quad X_{ij} = \delta_{ij} B.$$

Assume that

$$\rho = (\rho_{ij})_{i,j=1}^m = \sum_{i,j=1}^m E_{ij} \otimes \rho_{ij}$$

is a density matrix of the composite system written in block-matrix form. Then

$$\operatorname{Tr}(A \otimes I)\rho = \sum_{i,j} A_{ij} \operatorname{Tr} I_n \rho_{ij} = \sum_{i,j} A_{ij} \operatorname{Tr} \rho_{ij}$$

and this gives that for the first reduced density matrix ρ_1, we have

$$(\rho_1)_{ij} = \operatorname{Tr} \rho_{ij}. \tag{2.12}$$

We can compute similarly the second reduced density ρ_2. Since

$$\operatorname{Tr}(I \otimes B)\rho = \sum_i \operatorname{Tr} B \rho_{ii}$$

we obtain

$$\rho_2 = \sum_{i=1}^m \rho_{ii}. \tag{2.13}$$

The reduced density matrices might be expressed by the **partial traces**. The mappings $\operatorname{Tr}_2 : B(\mathcal{H}_1) \otimes B(\mathcal{H}_2) \to B(\mathcal{H}_1)$ and $\operatorname{Tr}_1 : B(\mathcal{H}_1) \otimes B(\mathcal{H}_2) \to B(\mathcal{H}_2)$ are defined as

$$\operatorname{Tr}_2(A \otimes B) = A \operatorname{Tr} B, \qquad \operatorname{Tr}_1(A \otimes B) = (\operatorname{Tr} A)B. \tag{2.14}$$

We have

$$\rho_1 = \operatorname{Tr}_2 \rho \qquad \text{and} \qquad \rho_2 = \operatorname{Tr}_1 \rho. \tag{2.15}$$

Axiom (A4) tells about a composite quantum system consisting of two quantum components. In case of more quantum components, the formalism is similar, but more tensor factors appear.

It may happen that the quantum system under study has a classical and a quantum component; assume that the first component is classical. Then the description by tensor product Hilbert space is still possible. A basis $(|e_i\rangle)_i$ of $\mathcal{H}_1$ can be fixed and the possible density matrices of the joint system are of the form

$$\sum_i p_i |e_i\rangle\langle e_i| \otimes \rho_i^{(2)}, \tag{2.16}$$

where $(p_i)_i$ is a probability distribution and $\rho_i^{(2)}$ are densities on $\mathcal{H}_2$. Then the reduced state on the first component is the probability density $(p_i)_i$ (which may be regarded as a diagonal density matrix) and $\sum_i p_i \rho_i^{(2)}$ is the second reduced density.

The next postulate of quantum mechanics tells about the **time development** of a closed quantum system. If the system is not subject to any measurement in the time interval $I \subset \mathbb{R}$ and ρ_t denotes the statistical operator at time t, then

(A5) $\rho_t = U(t,s)\rho_s U(t,s)^*$ $(t,s \in I)$,

where the **unitary propagator** $U(t,s)$ is a family of unitary operators such that

(i) $U(t,s)U(s,r) = U(t,r)$,
(ii) $(s,t) \mapsto U(s,t) \in B(\mathcal{H})$ is strongly continuous.

The first-order approximation of the unitary $U(s,t)$ is the **Hamiltonian**:

$$U(t+\Delta t, t) = I - \frac{i}{\hbar} H(t) \Delta t,$$

where $H(t)$ is the Hamiltonian at time t. If the Hamiltonian is time independent, then

$$U(s,t) = \exp\left(-\frac{i}{\hbar}(s-t)H\right).$$

In the approach followed here the density matrices are transformed in time, and this is the so-called **Schrödinger picture** of quantum mechanics. When discrete time development is considered, a single unitary U gives the transformation of the vector state in the form $\psi \mapsto U\psi$, or in the density matrix formalism $\rho \mapsto U\rho U^*$.

Example 2.7. Let $|0\rangle, |1\rangle, \ldots, |n-1\rangle$ be an orthonormal basis in an n-dimensional Hilbert space. The transformation

$$V : |i\rangle \mapsto \frac{1}{\sqrt{n}} \sum_{j=0}^{n-1} \omega^{ij} |j\rangle \qquad (\omega = e^{2\pi i/n}) \tag{2.17}$$

is a unitary and it is called **quantum Fourier transform**. $\qquad\qquad\square$

When the unitary time development is viewed as a quantum algorithm in connection with quantum computation, the term **gate** is used instead of unitary.

Example 2.8. Unitary operators are also used to manipulate quantum registers and to implement quantum algorithms.

The **Hadamard gate** is the unitary operator

$$U_H := \frac{1}{\sqrt{2}} \begin{bmatrix} 1 & 1 \\ 1 & -1 \end{bmatrix}. \tag{2.18}$$

It sends the basis vectors into uniform superposition and vice versa. The Hadamard gate can establish or destroy the superposition of a qubit. This means that the basis vector $|0\rangle$ is transformed into the vector $(|0\rangle + |1\rangle)/\sqrt{2}$, which is a superposition, and superposition is created.

The **controlled-NOT gate** is a unitary acting on two qubits. The first qubit is called "a control qubit," and the second qubit is the data qubit. This operator sends the basis vectors $|00\rangle, |01\rangle, |10\rangle, |11\rangle$ of $\mathbb{C}^4$ into $|00\rangle, |01\rangle, |11\rangle, |10\rangle$. When the first character is 1, the second changes under the operation. Therefore, the matrix of the controlled-NOT gate is

$$U_{c-NOT} := \begin{bmatrix} 1 & 0 & 0 & 0 \\ 0 & 1 & 0 & 0 \\ 0 & 0 & 0 & 1 \\ 0 & 0 & 1 & 0 \end{bmatrix}. \tag{2.19}$$

Fig. 2.2 The unitary made of the Hadamard gate, and the controlled-NOT gate transforms the standard product basis into the Bell basis

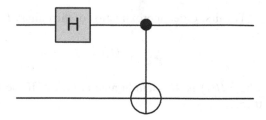

The **swap gate** moves a product vector $|i\rangle \otimes |j\rangle$ into $|j\rangle \otimes |i\rangle$. Therefore its matrix is

$$\begin{bmatrix} 1 & 0 & 0 & 0 \\ 0 & 0 & 1 & 0 \\ 0 & 1 & 0 & 0 \\ 0 & 0 & 0 & 1 \end{bmatrix}. \tag{2.20}$$

Quantum algorithms involve several other gates.   □

Example 2.9. The unitary operators are used to transform a basis into another one. In the Hilbert space $\mathbb{C}^4 = \mathbb{C}^2 \otimes \mathbb{C}^2$ the standard basis is

$$|00\rangle, |01\rangle, |10\rangle, |11\rangle.$$

The unitary

$$(U_H \otimes I_2) U_{c-NOT} = \frac{1}{\sqrt{2}} \begin{bmatrix} 1 & 0 & 1 & 0 \\ 0 & 1 & 0 & 1 \\ 0 & 1 & 0 & -1 \\ 1 & 0 & -1 & 0 \end{bmatrix}.$$

moves the standard basis into the so-called **Bell basis**:

$$\frac{1}{\sqrt{2}}(|00\rangle + |11\rangle), \quad \frac{1}{\sqrt{2}}(|01\rangle + |10\rangle), \quad \frac{1}{\sqrt{2}}(|00\rangle - |11\rangle), \quad \frac{1}{\sqrt{2}}(|01\rangle + |10\rangle).$$

This basis is complementary to the standard product basis (Fig. 2.2).

2.2 State Transformations

Assume that $\mathscr{H}$ is the Hilbert space of our quantum system which initially has a statistical operator ρ (acting on $\mathscr{H}$). When the quantum system is not closed, it is coupled to another system called **environment**. The environment has a Hilbert space $\mathscr{H}_e$ and statistical operator ρ_e. Before interaction the total system has density $\rho_e \otimes \rho$. The dynamical change caused by the interaction is implemented by a unitary, and $U(\rho_e \otimes \rho)U^*$ is the new statistical operator and the reduced density $\tilde{\rho}$ is the new statistical operator of the quantum system we are interested in. The affine change $\rho \mapsto \tilde{\rho}$ is typical for quantum mechanics and is called **state transformation**. In this way the map $\rho \mapsto \tilde{\rho}$ is defined on density matrices but it can be extended by linearity to all matrices. In this way we can obtain a trace-preserving and positivity preserving linear transformation.

The above-defined state transformation can be described in several other forms, and reference to the environment could be omitted completely. Assume that ρ is an $n \times n$ matrix and ρ_e is of the form $(z_k \overline{z_l})_{kl}$, where $(z_1, z_2, \ldots, z_m)$ is a unit vector in the m-dimensional space $\mathcal{H}_e$ (ρ_e is a pure state). All operators acting on $\mathcal{H}_e \otimes \mathcal{H}$ are written in a block-matrix form; they are $m \times m$ matrices with $n \times n$ matrix entries. In particular, $U = (U_{ij})_{i,j=1}^m$ and $U_{ij} \in M_n$. If U is a unitary, then U^*U is the identity and this implies that

$$\sum_i U_{ik}^* U_{il} = \delta_{kl} I_n \tag{2.21}$$

Formula (2.13) for the reduced density matrix gives

$$\tilde{\rho} = \mathrm{Tr}_1(U(\rho_e \otimes \rho)U^*) = \sum_i (U(\rho_e \otimes \rho)U^*)_{ii} = \sum_{i,k,l} U_{ik}(\rho_e \otimes \rho)_{kl}(U^*)_{li}$$

$$= \sum_{i,k,l} U_{ik}(z_k \overline{z_l} \rho)(U_{il})^* = \sum_i \left(\sum_k z_k U_{ik}\right) \rho \left(\sum_l z_l U_{il}\right)^* = \sum_i A_i \rho A_i^*,$$

where the operators $A_i := \sum_k z_k U_{ik}$ satisfy

$$\sum_p A_p^* A_p = I \tag{2.22}$$

in accordance with (2.21) and $\sum_k |z_k|^2 = 1$.

Theorem 2.1. *Any state transformation $\rho \mapsto \mathscr{E}(\rho)$ can be written in the form*

$$\mathscr{E}(\rho) = \sum_p A_p \rho A_p^*,$$

where the operator coefficients satisfy (2.22). Conversely, all linear mappings of this form are state transformations.

The first part of the theorem was obtained above. To prove the converse part, we need to solve the equations

$$A_i := \sum_k z_k U_{ik} \qquad (i = 1, 2, \ldots, m).$$

Choose simply $z_1 = 1$ and $z_2 = z_3 = \ldots = z_m = 0$ and the equations reduce to $U_{p1} = A_p$. This means that the first column is given from the block-matrix U and we need to determine the other columns in such a way that U should be a unitary. Thanks to the condition (2.22) this is possible. Condition (2.22) tells us that the first column of our block-matrix determines an isometry which extends to a unitary. $\qquad \square$

The coefficients A_p in the **operator-sum representation** are called the **operation elements** of the state transformation. The terms quantum (state) operation and channeling transformation are also often used instead of state transformation.

The state transformations form a convex subset of the set of all positive trace–preserving linear transformations. (It is not known what the extreme points of this set are.)

A linear mapping $\mathscr{E}$ is called **completely positive** if $\mathscr{E} \otimes id_n$ is positivity preserving for the identical mapping $id_n : M_n(\mathbb{C}) \to M_n(\mathbb{C})$ on any matrix algebra.

Theorem 2.2. *Let* $\mathscr{E} : M_n(\mathbb{C}) \to M_k(\mathbb{C})$ *be a linear mapping. Then* $\mathscr{E}$ *is completely positive if and only if it admits a representation*

$$\mathscr{E}(A) = \sum_u V_u A V_u^* \tag{2.23}$$

by means of some linear operators $V_u : \mathbb{C}^n \to \mathbb{C}^k$.

This result was first proven by Kraus. (Its proof and more detailed discussion of completely positive maps will be presented in the Appendix.) It follows that a state transformation is completely positive and the operator-sum representation is also called **Kraus representation**. Note that this representation is not unique.

Let $\mathscr{E} : M_n(\mathbb{C}) \to M_k(\mathbb{C})$ be a linear mapping. $\mathscr{E}$ is determined by the block-matrix $(X_{ij})_{1 \le i,j \le k}$, where

$$X_{ij} = \mathscr{E}(E_{ij}) \tag{2.24}$$

(Here E_{ij} denote the matrix units.) This is the **block-matrix representation** of $\mathscr{E}$.

Theorem 2.3. *Let* $\mathscr{E} : M_n(\mathbb{C}) \to M_k(\mathbb{C})$ *be a linear mapping. Then* $\mathscr{E}$ *is completely positive if and only if the representing block-matrix* $(X_{ij})_{1 \le i,j \le k} \in M_k(\mathbb{C}) \otimes M_n(\mathbb{C})$ *is positive.*

Example 2.10. Consider the transpose mapping $A \mapsto A^T$ on 2×2 matrices:

$$\begin{bmatrix} x & y \\ z & w \end{bmatrix} \mapsto \begin{bmatrix} x & z \\ y & w \end{bmatrix}.$$

The representing block-matrix is

$$X = \begin{bmatrix} 1 & 0 & 0 & 0 \\ 0 & 0 & 1 & 0 \\ 0 & 1 & 0 & 0 \\ 0 & 0 & 0 & 1 \end{bmatrix}.$$

This is not positive, so the transpose mapping is not completely positive.   □

Example 2.11. Consider a positive trace–preserving transformation $\mathscr{E} : M_n(\mathbb{C}) \to M_m(\mathbb{C})$ such that its range consists of commuting operators. We can show that $\mathscr{E}$ is automatically a state transformation.

Since a commutative subalgebra of $M_m(\mathbb{C})$ is the linear span of some pairwise orthogonal projections P_k, one can see that $\mathscr{E}$ has the form

$$\mathscr{E}(A) = \sum_k P_k \operatorname{Tr} F_k A, \qquad (2.25)$$

where F_k is a positive operator in $M_n(\mathbb{C})$; it induces the coefficient of P_k as a linear functional on $M_n(\mathbb{C})$.

The positivity of the representing block-matrix is

$$\sum_{ij} E_{ij} \otimes \left(\sum_k P_k \operatorname{Tr}(F_k E_{ij}) \right) = \sum_k \left(\sum_{ij} E_{ij} \otimes P_k \right) \circ \left(\sum_{ij} E_{ij} \operatorname{Tr}(F_k E_{ij}) \otimes I \right),$$

where $\circ$ denotes the Hadamard (or entry-wise product) of $nm \times nm$ matrices. Recall that according to Schur's theorem the **Hadamard product** of positive matrices is positive. The first factor is

$$[P_k, P_k, \ldots, P_k]^* [P_k, P_k, \ldots, P_k]$$

and the second factor is $F_k \otimes I$; both are positive.

Consider the particular case of (2.25) where each P_k is of rank one and $\sum_{k=1}^r F_k = I$. Such a family of F_k's describe a measurement which associates the r-tuple $(\operatorname{Tr} \rho F_1, \operatorname{Tr} \rho F_2, \ldots, \operatorname{Tr} \rho F_r)$ to the density matrix ρ. Therefore a measurement can be formulated as a state transformation with diagonal outputs. $\qquad \square$

The Kraus representation and the block-matrix representation are convenient ways to describe a state transformation in any finite dimension. In the 2×2 case we have the possibility to expand the mappings in the basis $\sigma_0, \sigma_1, \sigma_2, \sigma_3$.

Any trace-preserving mapping $\mathscr{E} : M_2(\mathbb{C}) \to M_2(\mathbb{C})$ has a matrix

$$T = \begin{bmatrix} 1 & 0 \\ t & T_3 \end{bmatrix} \qquad (2.26)$$

with respect to this basis, where $T_3 \in M_3$ and

$$\mathscr{E}(w_0 \sigma_0 + w \cdot \sigma) = w_0 \sigma_0 + (t + T_3 w) \cdot \sigma. \qquad (2.27)$$

Since $\mathscr{E}$ sends self-adjoint operators to self-adjoint operators, we may assume that T_3 is a real 3×3 matrix. It has a singular value decomposition $O_1 \Sigma O_2$, where O_1 and O_2 are orthogonal matrices and Σ is diagonal. Since any orthogonal transformation on $\mathbb{R}^3$ is induced by a unitary conjugation on $M_2(\mathbb{C})$, in the study of state transformations we can assume that T_3 is diagonal.

The following examples of state transformations are given in terms of the T-representation.

Example 2.12 (Pauli channels). $t = 0$ and $T_3 = \operatorname{Diag}(\alpha, \beta, \gamma)$. Density matrices are sent to density matrices if and only if

$$-1 \leq \alpha, \beta, \gamma \leq 1$$

for the real parameters α, β, γ.

It is not difficult to compute the representing block-matrix, we have

$$X = \begin{bmatrix} \frac{1+\gamma}{2} & 0 & 0 & \frac{\alpha+\beta}{2} \\ 0 & \frac{1-\gamma}{2} & \frac{\alpha-\beta}{2} & 0 \\ 0 & \frac{\alpha-\beta}{2} & \frac{1-\gamma}{2} & 0 \\ \frac{\alpha+\beta}{2} & 0 & 0 & \frac{1+\gamma}{2} \end{bmatrix}. \tag{2.28}$$

X is unitarily equivalent to the matrix

$$\begin{bmatrix} \frac{1+\gamma}{2} & \frac{\alpha+\beta}{2} & 0 & 0 \\ \frac{\alpha+\beta}{2} & \frac{1+\gamma}{2} & 0 & 0 \\ 0 & 0 & \frac{1-\gamma}{2} & \frac{\alpha-\beta}{2} \\ 0 & 0 & \frac{\alpha-\beta}{2} & \frac{1-\gamma}{2} \end{bmatrix}.$$

This matrix is obviously positive if and only if

$$|1 \pm \gamma| \geq |\alpha \pm \beta|. \tag{2.29}$$

This positivity condition holds when $\alpha = \beta = \gamma = p > 0$. Hence the next example gives a channeling transformation.   □

Example 2.13 (Depolarizing channel). This channel is given by the matrix T from (2.26), where $t = 0$ and $T_3 = pI$. Assume that $0 < p < 1$.

Since

$$\mathscr{E}_p(\tfrac{1}{2}\sigma_0 + w \cdot \sigma) = p(\tfrac{1}{2}\sigma_0 + w \cdot \sigma) + (1-p)\tfrac{1}{2}\sigma_0 = \tfrac{1}{2}\sigma_0 + p(w \cdot \sigma),$$

the depolarizing channel keeps the density with probability p and moves to the completely apolar state $\sigma_0/2$ with probability $1 - p$.

Extension to n-level system is rather obvious. $\mathscr{E}_{p,n} : M_n \to M_n$ is defined as

$$\mathscr{E}_{p,n}(A) = pA + (1-p)\frac{I}{n}\mathrm{Tr}\,A. \tag{2.30}$$

is trivially completely positive for $0 \leq p \leq 1$, since it is the convex combination of such mappings. In order to consider the negative values of p we should study the representing block-matrix X. One can see that

$$X = p\sum_{ij} E_{ij} \otimes E_{ij} + \frac{1-p}{n}I \otimes I.$$

The matrix $\frac{1}{n}\sum_{ij} E_{ij} \otimes E_{ij}$ is a self-adjoint idempotent (that is, a projection), so its spectrum is $\{0, 1\}$. Consequently, the eigenvalues of X are

$$pn + \frac{1-p}{n}, \frac{1-p}{n}.$$

They are positive when

$$-\frac{1}{n^2 - 1} \le p \le 1. \tag{2.31}$$

This is the necessary and sufficient condition for the complete positivity of $\mathscr{E}_{p,n}$. $\square$

Example 2.14 (Phase-damping channel). $t = 0$ and $T_3 = \mathrm{Diag}(p, p, 2p - 1)$. This channel describes decoherence, the decay of a superposition into a mixture;

$$\mathscr{E} \begin{bmatrix} a & b \\ \bar{b} & c \end{bmatrix} = (1 - p) \begin{bmatrix} a & b \\ \bar{b} & c \end{bmatrix} + p \begin{bmatrix} a & 0 \\ 0 & c \end{bmatrix}.$$

$\square$

Example 2.15 (Fuchs channel). This channel is not unit preserving and maps σ_2 into 0:

$$T = \begin{bmatrix} 1 & 0 & 0 & 0 \\ 0 & \dfrac{1}{\sqrt{3}} & 0 & 0 \\ 0 & 0 & 0 & 0 \\ \dfrac{1}{3} & 0 & 0 & \dfrac{1}{3} \end{bmatrix}$$

The Fuchs channel is an extreme point in the convex set of channels $M_2(\mathbb{C}) \to M_2(\mathbb{C})$. Figure 2.3 is an illustration. $\square$

Example 2.16 (Amplitude-damping channel).

$$T = \begin{bmatrix} 1 & 0 & 0 & 0 \\ 0 & \sqrt{1-p} & 0 & 0 \\ 0 & 0 & \sqrt{1-p} & 0 \\ p & 0 & 0 & 1-p \end{bmatrix}$$

or equivalently

$$\mathscr{E} \begin{bmatrix} a & b \\ \bar{b} & c \end{bmatrix} = \begin{bmatrix} a + pc & \sqrt{1-p}\,b \\ \sqrt{1-p}\,\bar{b} & (1-p)c \end{bmatrix}.$$

The Bloch ball shrinks toward the north pole (Fig. 2.4). $\square$

Example 2.17 (The Holevo–Werner channel). Set a linear mapping $\mathscr{E} : M_n \to M_n$ as

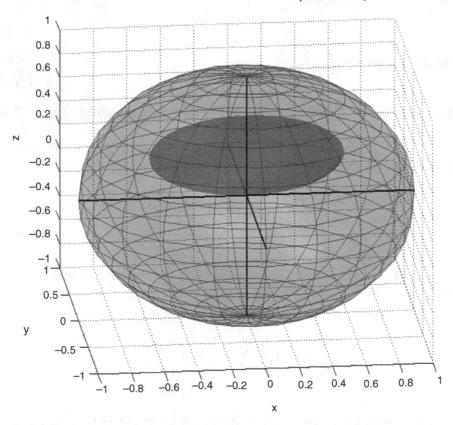

Fig. 2.3 The Fuchs channel maps the Bloch ball into an ellipse in the $x - z$ plane

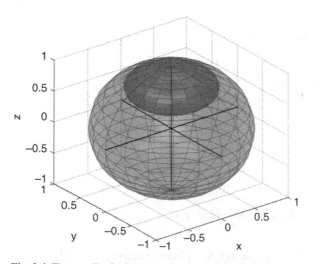

Fig. 2.4 The amplitude-damping channel shrinks the Bloch ball toward the north pole

$$\mathscr{E}(D) = \frac{1}{n-1}(\mathrm{Tr}\,(D)I - D^T),$$

where D^T denotes the transpose of D. The positivity is not obvious from this form but it is easy to show that

$$\mathscr{E}(D) = \frac{1}{2(n-1)} \sum_{i,j}(E_{ij} - E_{ji})^* D(E_{ij} - E_{ji}),$$

where E_{ij} denotes the matrix units. This is the Kraus representation of $\mathscr{E}$ which must be completely positive.

In the space of matrices the following matrices are linearly independent.

$$d_k = \mathrm{Diag}\,(\overset{1}{1}, \overset{2}{1}, \ldots, \overset{n-k}{1}, -(n-k), 0, 0, \ldots, 0).$$

For $k = 0$ we have the unit matrix and d_1, d_2, d_{n-1}, are traceless matrices. Moreover, set

$$e_{ij} = E_{ij} - E_{ji} \qquad (1 \le i < j \le n),$$
$$f_{ij} = -iE_{ij} + iE_{ji} \qquad (1 \le i < j \le n).$$

The matrices $\{d_k : 0 \le k \le n-1\} \cup \{e_{ij} : 1 \le i < j \le n\} \cup \{f_{ij} : 1 \le i < j \le n\}$ are pairwise orthogonal with respect to the Hilbert Schmidt inner product, and up to a normalizing factor they form a basis in the Hilbert space M_n. (Actually these matrices are a kind of generalization of the Pauli matrices for $n > 2$.)

The mapping $\mathscr{E}$ is unital, hence $\mathscr{E}(d_0) = d_0$. For $0 < k < n$ we have

$$\mathscr{E}(d_k) = \frac{1}{n-1}d_k$$

and for $1 \le i < j \le n$ we have

$$\mathscr{E}(E_{ij}) = E_{ij}$$
$$\mathscr{E}(F_{ij}) = -F_{ij}.$$

Hence our basis consists of eigenvectors, the spectrum of $\mathscr{E}$ is $\{1, -1, \frac{1}{n-1}\}$ with the multiplicities $n\,(n-1)/2 + 1$, $n(n-1)/2$, $n-2$, respectively. Although $\mathscr{E}$ is completely positive, its spectrum contains negative numbers; therefore it is not true that $\mathscr{E}$ is positive definite with respect to the Hilbert–Schmidt inner product. $\qquad\square$

Example 2.18 (Transpose depolarizing channel). Let $\mathscr{E}_{p,n}^T : M_n \to M_n$ is defined as

$$\mathscr{E}_{p,n}^T(A) = tA^T + (1-t)\frac{I}{n}\mathrm{Tr}A, \qquad (2.32)$$

where A^T is the transpose of A. In order to decide the complete positivity, we should study the representing block-matrix X:

$$X = t \sum_{ij} E_{ij} \otimes E_{ji} + \frac{1-t}{n} I \otimes I.$$

The matrix $\sum_{ij} E_{ij} \otimes E_{ji}$ is self-adjoint and its square is the identity. Therefore, its spectrum is $\{\pm 1\}$. Consequently, the eigenvalues of X are

$$-t + \frac{1-p}{n}, p + \frac{1-t}{n}.$$

They are positive when

$$-\frac{1}{n-1} \le t \le \frac{1}{n+1}. \tag{2.33}$$

This is the necessary and sufficient condition for the complete positivity of $\mathscr{E}^T_{p,n}$. The Holevo–Werner channel is a particular case.   □

2.3 Notes

There are several books about the mathematical foundations of quantum mechanics. The book of von Neumann [78] has a historical value; it was published in 1932. Holevo's lecture note [56] is rather comprehensive and [19] treats unbounded linear operators of Hilbert spaces in detail.

The axioms of quantum mechanics are not so strict as in mathematics.

2.4 Exercises

1. Show that the vectors $|x_1\rangle, |x_2,\rangle, \ldots, |x_n\rangle$ form an orthonormal basis in an n-dimensional Hilbert space if and only if

$$\sum_i |x_i\rangle\langle x_i| = I.$$

2. Express the Pauli matrices in terms of the ket vectors $|0\rangle$ and $|1\rangle$.
3. Show that the Pauli matrices are unitarily equivalent.
4. Show that for a 2×2 matrix A the relation

$$\frac{1}{2} \sum_{i=0}^{3} \sigma_i A \sigma_i = (\mathrm{Tr} A) I$$

   holds.
5. Let t be a real number and n be a unit vector in $\mathbb{R}^3$. Show that

$$\exp(itn \cdot \sigma) = \cos t \, (n \cdot \sigma) + i \sin t \, (n \cdot \sigma).$$

6. Let v and w be complex numbers and let n be a unit vector in $\mathbb{R}^3$. Show that

$$\exp(v\sigma_0 + w(n \cdot \sigma)) = e^v \left((\cosh w) \, \sigma_0 + (\sinh w) n \cdot \sigma\right). \tag{2.34}$$

7. Let $|1\rangle, |2\rangle, |3\rangle$ be a basis in $\mathbb{C}^3$ and

$$|v_1\rangle = \frac{1}{\sqrt{3}}(|1\rangle + |2\rangle + |3\rangle),$$

$$|v_2\rangle = \frac{1}{\sqrt{3}}(|1\rangle - |2\rangle - |3\rangle),$$

$$|v_3\rangle = \frac{1}{\sqrt{3}}(|1\rangle - |2\rangle + |3\rangle),$$

$$|v_4\rangle = \frac{1}{\sqrt{3}}(|1\rangle + |2\rangle - |3\rangle).$$

Compute $\sum_{i=1}^{4} |v_i\rangle\langle v_i|$.

8. Show the identity

$$(I_2 - \sigma_k) \otimes (I_2 + \sigma_k) + (I_2 + \sigma_k) \otimes (I_2 - \sigma_k) = I_4 - \sigma_k \otimes \sigma_k \qquad (2.35)$$

for $k = 1, 2, 3$.

9. Let e and f be unit vectors in $\mathbb{C}^2$ and assume that they are eigenvectors of two different Pauli matrices. Show that

$$|\langle e, f\rangle|^2 = \frac{1}{2}.$$

10. Consider two complementary orthonormal bases in a Hilbert space. Let A and B operators be such that $\text{Tr}\, A = \text{Tr}\, B = 0$; A is diagonal in the first basis, while B is diagonal in the second one. Show that $\text{Tr}\, AB = 0$.

11. Show that the number of pairwise **complementary orthonormal bases** in an n-dimensional Hilbert space is at most $n + 1$. (Hint: Estimate the dimension of the subspace of traceless operators.)

12. Show that the quantum Fourier transform moves the standard basis to a complementary basis.

13. What is the 8×8 matrix of the controlled-swap gate? (This unitary is called **Fredkin gate**.)

14. Give the density matrix corresponding to the singlet state (2.9) and compute the reduced density matrices.

15. Let ρ be a density matrix. Show that ρ corresponds to a pure state if and only if $\rho^2 = \rho$.

16. Compute the dimension of the set of the extreme points and the dimension of the topological boundary of the $n \times n$ density matrices.

17. Compute the reduced density matrices of the state

$$\frac{1}{3}\begin{bmatrix} 1 & 0 & 0 & 0 \\ 0 & 1 & 1 & 0 \\ 0 & 1 & 1 & 0 \\ 0 & 0 & 0 & 0 \end{bmatrix}.$$

18. Let $0 < p < 1$. Show that the Kraus representation of the depolarizing channel on M_2 is

$$\mathscr{E}_{p,2}(A) = \frac{3p+1}{4}A + \frac{1-p}{4}\sigma_1 A \sigma_1 + \frac{1-p}{4}\sigma_2 A \sigma_2 + \frac{1-p}{4}\sigma_3 A \sigma_3.$$

19. Assume that $\mathscr{E} : M_n(\mathbb{C}) \to M_n(\mathbb{C})$ is defined as

$$\mathscr{E}(A) = \frac{1}{n-1}(I \operatorname{Tr} A - A).$$

Show that $\mathscr{E}$ is positive but not completely positive.

20. Let p be a real number. Show that the mapping $\mathscr{E}_{p,2} : M_2 \to M_2$ is defined as

$$\mathscr{E}_{p,2}(A) = pA + (1-p)\frac{I}{2}\operatorname{Tr} A$$

is positive if and only if $-1 \le p \le 1$. Show that $\mathscr{E}_{p,2}$ is completely positive if and only if $-1/3 \le p \le 1$. (Hint: $\mathscr{E}_{p,2}$ is a Pauli channel.)

21. Let $\mathscr{E}_{p,2} : M_2 \to M_2$ be the depolarizing channel and A^T is the transpose of A. For which values of the parameter p will $A \mapsto \mathscr{E}_{p,2}(A)^T$ be a state transformation?

22. What is the spectrum of the linear mapping $\mathscr{E}_{p,2}$? Can a positive mapping have negative eigenvalues?

23. Give the Kraus representation of the phase-dumping channel.

24. Show that

$$\frac{1}{3}\begin{bmatrix} 5 & 0 & 0 & \sqrt{3} \\ 0 & 1 & i\sqrt{3} & 0 \\ 0 & -i\sqrt{3} & 3 & 0 \\ \sqrt{3} & 0 & 0 & 3 \end{bmatrix}$$

is the representing block-matrix of the Fuchs channel.

25. Show that the matrix

$$T = \begin{bmatrix} 1 & 0 & 0 & 0 \\ 0 & \cos\delta & 0 & 0 \\ 0 & 0 & \cos\gamma & 0 \\ \sin\gamma\sin\delta & 0 & 0 & \cos\gamma\cos\delta \end{bmatrix}$$

determines a state transformation of a qubit.

26. Compute the limit of $\mathscr{E}^n$ if $\mathscr{E}$ is the amplitude-damping channel.

27. Assume that $\mathscr{E} : M_n(\mathbb{C}) \to M_n(\mathbb{C})$ acts as

$$\mathscr{E}(A)_{ij} = \delta_{ij}A_{ij},$$

that is, $\mathscr{E}$ kills all off-diagonal entries. Find the Kraus representation of $\mathscr{E}$.

Chapter 3
Information and its Measures

Information must be written on some physical substance, which could be the neural connections in our brain, a piece of paper, some magnetic media, electrons trapped in quantum dots, a beam of photons, and there are many other possibilities. Shortly speaking, *"information is physical."* Physics of the media that is used to store information determines the way how to manipulate that information. Classical media, for example transistors, magnetic domains, or paper, determine classical logic as the means to manipulate that information. In classical logic things are true or false; for example, a magnetic domain on the drive either is aligned with the direction of the head or is not. Any memory location can be read without destroying that memory location.

Given a set $\mathscr{X}$ of outcomes of an experiment, information gives one of the possible alternatives. The value (or measure) of the information is proportional to the size of $\mathscr{X}$. The idea of measuring information regardless of its content dates back to R.V.L. Hartley (1928). Hartley made an attempt at the determination of an information measure and he recognized the logarithmic nature of the natural measure of information content: When the cardinality of $\mathscr{X}$ is n, the amount of information assigned to an element is $\log n$. In this spirit, the information content of a word of a dictionary of n words equals to $\log n$. Actually, Hartley took the logarithm to the base 10, but a different base yields only an extra constant factor. The choice of base 2 could be more natural. Suppose that $\mathscr{X}$ is the set of natural numbers $\{1, 2, \ldots, n\}$ and they are written in a binary way in the form of a 0–1 sequence. Then we need $\lceil \log_2 n \rceil$ binary digits to describe an outcome. (Here $\lceil \log_2 n \rceil$ denotes the smallest integer which is not smaller than $\log_2 n$.)

Quantum information is carried by a physical system obeying the laws of quantum mechanics. Any measurement performed on a quantum system destroys most of the information contained in that system and leaves the system in one of the so-called basis states. The discarded information is unrecoverable. The outcome of the measurement is stochastic; in general, only probabilistic predictions can be made. Quantum information can be transferred with perfect fidelity, but in the process the original must be destroyed (see Chap. 4). The quantum teleportation protocol was

D. Petz, *Information and its Measures*. In: D. Petz, Quantum Information Theory and Quantum Statistics, Theoretical and Mathematical Physics, pp. 25–51 (2008)
DOI 10.1007/978-3-540-74636-2_3

first described only in 1993, in spite of the fact that all the physical background was known already in the 1920s. (An experimental demonstration was carried out in 1998.)

3.1 Shannon's Approach

In his revolutionary paper Shannon proposed a statistical approach and he posed the problem in the following way: *"Suppose we have a set of possible events whose probabilities of occurrence are p_1, p_2, ..., p_n. These probabilities are known but that is all we know concerning which event will occur. Can we find a measure of how much "choice" is involved in the selection of the event or how uncertain we are of the outcome?"* Denoting such a measure by $H(p_1, p_2, ..., p_n)$, he listed three very reasonable requirements which should be satisfied. He concluded that the only H satisfying the three assumptions is of the form

$$H(p_1, p_2, ..., p_n) = -K \sum_{i=1}^{n} p_i \log p_i, \tag{3.1}$$

where K is a positive constant. Nowadays we call this quantity **Shannon entropy**. It is often said that Shannon's proofs were rather sketchy. His postulational approach to the amount of information was improved by others and the following axioms are standard today:

(1) **Continuity**: $H(p, 1-p)$ is a continuous function of p.
(2) **Symmetry**: $H(p_1, p_2, ..., p_n)$ is a symmetric function of its variables.
(3) **Recursion**: For every $0 \le \lambda < 1$ the recursion $H(p_1, ..., p_{n-1}, \lambda p_n, (1-\lambda)p_n) = H(p_1, ..., p_n) + p_n H(\lambda, 1-\lambda)$ holds.

Beyond the postulational approach to the information measure, the following argument provides a justification of the quantity H. Assume that the outcomes of an experiment show up with probabilities p_1, p_2, ..., p_n; the experiment is performed but the actual outcome is recorded by another person who is willing to answer affirmative questions (answered by "yes" or "no"). If one aims to learn the result of the experiment by interrogating the observer, the number of necessary questions depends on both the probabilities p_1, p_2, ..., p_n and the questioning strategy. Suppose that the interrogator wants to minimize the number of questions on the basis of the probability distribution known to him. He divides the outcomes into two disjoint subsets A and B, so that the probability of A and B are approximately the same, that is, about one half. A good first question is to ask whether the outcome belongs to A. In this way the interrogator can strike out half of the possibilities (measured in probability) regardless of the answer "yes" or "no". After the first question, the interrogator divides the remaining possibilities into two parts of equal probability, etc. The expected number of questions is the Shannon entropy, when this strategy is followed. (The constant K is chosen to be 1 and logarithm is based to 2.)

Since the Shannon entropy is mostly a non-integer and the division of the possibilities into events of equal probability is usually not possible, the above scheme must be slightly refined. Assume that the experiment is performed N times under independent circumstances and the outcome is recorded by the same observer who is interrogated afterwards about all outcomes. In this way the interrogator can find events of probability 1/2 for the repeated observation more easily, but the number of necessary questions will increase with N. However, the expected number of questions divided by the number N of repetitions tends to the Shannon entropy. (The result explained in this interrogative scheme is the noiseless channel coding theorem. This theorem is independent of the axiomatic characterization of entropy and provides another interpretation of this quantity.)

The Shannon entropy is maximal on the uniform distribution:

$$H(p_1, p_2, \ldots, p_n) \leq \log n \tag{3.2}$$

This property follows easily from the concavity of the logarithm and has the interpretation that our uncertainty about the result of an experiment is maximal when all possible outcomes are equally probable.

Let X and Y be random variables with values in the sets $\mathscr{X}$ and $\mathscr{Y}$. The following notation will be used.

$$p(x) = \text{Prob}(X = x), \quad p(y) = \text{Prob}(Y = y),$$
$$p(x, y) = \text{Prob}(X = x, Y = y), \quad p(x|y) = \text{Prob}(X = x | Y = y).$$

If X_1, X_2, $\ldots$, X_k are random variables (with finite range), then the notation $H(X_1, X_2, \ldots, X_k)$ stands for the Shannon entropy of the joint distribution. In particular,

$$H(X, Y) := - \sum_{x \in \mathscr{X}} \sum_{y \in \mathscr{Y}} p(x, y) \log p(x, y). \tag{3.3}$$

The Shannon entropy is **subadditive**:

Theorem 3.1. *If X_1, X_2, $\ldots$, X_k are random variables of finite range, then*

$$H(X_1, X_2, \ldots, X_k) \leq \sum_{i=1}^{k} H(X_i).$$

The proof gives occasion to introduce the **conditional entropy**. With the above notation

$$H(X|Y) := - \sum_{y \in \mathscr{Y}} p(y) \sum_{x \in \mathscr{X}} p(x|y) \log p(x|y). \tag{3.4}$$

or equivalently

$$H(X|Y) := - \sum_{y \in \mathscr{Y}} \sum_{x \in \mathscr{X}} p(x, y) \log p(x|y). \tag{3.5}$$

The first equation shows that $H(X|Y)$ is the convex combination of some Shannon entropies, therefore $H(X|Y) \geq 0$.

To prove Theorem 3.1 (for $k = 2$), we may proceed as follows. First verify the **chain rule**

$$H(X,Y) = H(X) + H(Y|X) \tag{3.6}$$

and then use the property

$$H(X|Y) \leq H(X). \tag{3.7}$$

(Note that $H(X) - H(X|Y)$ is the mutual information, which is positive, see (5.3).)

A little modification of this argument yields a stronger result. By addition of the relations

$$H(X,Y,Z) = H(X,Y) + H(Z|X,Y)$$
$$H(Y) + H(Z|Y) = H(Y,Z)$$
$$H(Z|X,Y) \leq H(Z|Y)$$

we can conclude the **strong subadditivity**.

Theorem 3.2. *If X, Y and Z are random variables of finite range, then*

$$H(X,Y,Z) \leq H(X,Y) + H(Y,Z) - H(Y).$$

Note that the strong subadditivity may be written in the form

$$H(X|Y,Z) \leq H(X|Y). \tag{3.8}$$

A basic estimate for the conditional entropy is provided by the **Fano's inequality**:

Theorem 3.3. *Let X and Y be random variables such that their range is in a set of cardinality d and let $p := \mathrm{Prob}(X \neq Y)$. Then*

$$H(X|Y) \leq p \log(d-1) + H(p, 1-p).$$

3.2 Classical Source Coding

In this section the setting of the classical source coding is used to give motivation to Shannon's entropy and to the relative entropy.

Let X be a random variable with a finite range $\mathscr{X}$. A **source code** C for X is a mapping from $\mathscr{X}$ to the set of finite-length strings of symbols of a d-ary alphabet, which is assumed to be the set $\{0, 1, 2, \ldots, d-1\}$. Let $C(x)$ denote the codeword corresponding to x and let $\ell(x)$ denote the length of $C(x)$. If $p(x)$ is the probability of $x \in \mathscr{X}$, then the **expected length** of a source code C is given by

$$L(C) := \sum_{x} p(x)\ell(x).$$

Since the transmission of lengthy codewords could be costly, the aim of source coding is to make the expected code-length as small as possible. It is obvious that to meet this requirement the most frequent outcome of X must have the shortest codeword. For example, in the Morse code the letter e (which is the most frequent one in both English and Hungarian) is represented by a single dot. (The **Morse code** uses an alphabet of four symbols: a dot, a dash, a letter space and a word space.) The extension of a code C to the finite-length strings of $\mathscr{X}$ is defined by

$$C^*(x_1 x_2 \ldots x_n) = C(x_1)C(x_2)\ldots C(x_n),$$

where the right-hand side is the concatenation of the corresponding codewords.

A code C is **uniquely decodable** if $C^*(x_1 x_2 \ldots x_n) = C^*(x_1' x_2' \ldots x_m')$ implies that $x_1 x_2 \ldots x_n = x_1' x_2' \ldots x_m'$, that is, $n = m$ and $x_i = x_i'$ for all $1 \leq i \leq n$. A code is called **prefix code** if no codeword is a prefix of any other. In case of a prefix code the end of a codeword is immediately recognized and hence such a code is uniquely decodable. For example, if 0, 10, 110 and 111 are the binary codewords (of a prefix code), then the binary string 1011001101110 is easily decomposed into 6 codewords: 10,110,0,110,111,0.

Theorem 3.4 (Kraft–MacMillan). *The codeword lengths $\ell(x)$ of a uniquely decodable code over an alphabet of size d satisfy the inequality*

$$\sum_x d^{-\ell(x)} \leq 1.$$

Conversely, given a set of codeword lengths that satisfy this inequality, there exists a prefix code with these codewords lengths.

The proof is available in several standard books, for example [26]. It follows from the theorem that a uniquely decodable code could always be replaced by a prefix code which has the same codeword lengths.

Let $\lceil t \rceil$ denote the smallest integer $\geq t \in \mathbb{R}$. The codeword lengths $\ell(x) :=$ $\lceil -\log_d p(x) \rceil$ satisfy the Kraft inequality

$$\sum_x d^{-\ell(x)} \leq \sum_x p(x) = 1.$$

According to the theorem there exists a prefix code with these codeword lengths. (Such a code is called **Shannon code**.) Since $-\log_d p(x) \leq \ell(x) \leq -\log_d p(x) + 1$, we have

$$-\sum_x p(x) \log_d p(x) \leq L(C) \leq 1 - \sum_x p(x) \log_d p(x).$$

for the expected code-length $L(C)$. For the rest let us assume that $d = 2$. Then the bounds are given in terms of the **Shannon entropy** $H(p(x)) := -\sum_x p(x) \log p(x)$ as

$$H(p(x)) \leq L(C) \leq H(p(x)) + 1.$$

According to the next theorem the Shannon code is close to optimal.

Theorem 3.5. *The expected code-length of any prefix code is greater than or equal to the Shannon entropy of the source.*

Proof. $L - H(p(x)) \geq 0$ can be shown as follows:

$$
\begin{aligned}
L - H(p) &= \sum_x p(x)\ell(x) + \sum_x p(x)\log p(x) \\
&= -\sum_x p(x)\log 2^{-\ell(x)} + \sum_x p(x)\log p(x) \\
&= \sum_x p(x)\log \frac{p(x)}{r(x)} - \log c,
\end{aligned}
$$

where $r(x) = c^{-1}2^{-\ell(x)}$ and $c = \sum_x 2^{-\ell(x)}$. The **relative entropy** of two probability distributions is defined as

$$
D(p\|r) := \sum_x p(x)\big(\log p(x) - \log r(x)\big) \tag{3.9}
$$

and this quantity is known to be positive and 0 if and only if $p = q$. In terms of the relative entropy we have

$$
L - H(p) = D(p\|r) + \log\frac{1}{c}.
$$

Since $D(p\|r) \geq 0$ and $c \leq 1$ from the Kraft–McMillan inequality, this shows $L - H(p) \geq 0$. $\square$

The Shannon code is close to optimal only if we know correctly the distribution of the source X. Assume that it is not the case, and associate to x the codeword length $\lceil -\log q(x)\rceil$, where q is another probability distribution on $\mathscr{X}$, possibly different from the true distribution p. One can compute that in this case

$$
H(p) + D(p\|q) \leq L(C) \leq H(p) + D(p\|q) + 1. \tag{3.10}
$$

For the use of the wrong distribution the relative entropy is the penalty in the expected length.

The optimal coding is provided by a procedure due to Huffman. The **Huffman code** is not easy to describe, therefore I will show another coding due to Fano. The **Fano code** is nearly optimal, and it satisfies the inequality

$$
L(C) \leq H(p) + 2.
$$

In the Fano coding the probabilities $p(x)$ are ordered decreasingly as $p_1 \geq p_2 \geq p_3 \geq \ldots \geq p_m$. k is chosen such that

$$
\left| \sum_{i=1}^k p_i - \sum_{i=k+1}^m p_i \right|
$$

is minimal. The division of the probabilities into the two classes divides the source symbols into two classes. Assign 0 for the first bit of the lower class and 1 for the first bit of the upper class. The two classes have nearly equal probabilities. Then we can repeat the procedure for each of the two classes to determine the further bits of the code strings. This is Fano's scheme.

Up to now we have dealt with uniquely decodable codes. If the transmission of lengthy codewords is expensive, we might give up the exact decodability provided that the probability of mistake is small and long codewords can be avoided. This is a different approach to coding and decoding. Assume that the source emits the symbols X_1, X_2, X_3, ..., X_n (independently and according to the same distribution p, typical for the source). We can fix a coding procedure and all the emitted symbols are coded by this procedure, which could be the Fano code, for example. Let L_1, L_2, ..., L_n be the code-length of X_1, X_2, ...X_n, respectively. Both X_1, X_2, ..., X_n and L_1, L_2, ..., L_n are identically distributed independent random variables, and the expectation of L_i is $L(C)$. The law of large numbers tells us that the probability of the event

$$L_1 + L_2 + \cdots + L_n \geq n\big(L(C) + \varepsilon\big) \tag{3.11}$$

goes to 0 as $n \to \infty$. When x_1, x_2, ..., x_n is a string of source symbols such that the corresponding code string is shorter than $n(L(C) + \varepsilon)$, then we can code the string $x_1 x_2 \ldots x_n$ perfectly, otherwise we can use always the same code string. If the latter case happens to occur, then we cannot recover the emitted symbol string from the code string. However, the probability of this error is exactly the probability of the event (3.11), which tends to 0. What did we win in this way? The number of source strings is $|\mathscr{X}|^n$ and the number of binary strings used in the coding is $2^{n(L(C)+\varepsilon)}$. When $L(C) < \log|\mathscr{X}|$, then

$$2^{n(L(C)+\varepsilon)} \ll |\mathscr{X}|^n.$$

Hence the cardinality of our code book is much smaller than the cardinality of the source strings if a small probability of error is allowed. We can also say that the data set $\mathscr{X}^n$ is compressed to a set of binary strings of length $n(L(C)+\varepsilon)$. What we have is an example of **data compression**. Efficient data compression is the same as source coding by short binary code strings. Since we need $n(L(C)+\varepsilon)$ binary digits for a source string of length n, $L(C) + \varepsilon$ is called **code rate**. (It is the number of binary digits needed for a single source symbol, in the average.) Using the Shannon code, we can achieve a code rate $H(p) + \varepsilon$. However, if we mistake the distribution of the source and assume q instead of p, then the rate is higher; it is about $H(p) + D(p\|q) + \varepsilon$. Hence the above method is very sensitive for the distribution of the source. To avoid this and to achieve slightly better code rate **block coding** can be used. Shortly speaking, block coding means that the source string is not coded letter by letter but the whole string gets a code string.

A **block code** $(2^{nR_n}, n)$ for a source X_1, X_2, ..., X_n is given by two (sequences of) mappings:

$$f_n : \mathcal{X}^n \to \{1, 2, \ldots, 2^{nR_n}\}, \qquad \phi_n : \{1, 2, \ldots, 2^{nR_n}\} \to \mathcal{X}^n.$$

Here f_n is the **encoder**, ϕ_n is the **decoder** and $R := \lim R_n$ is called the **rate of the code**. The **probability of error** of the code is

$$P_e^{(n)} := \mathrm{Prob}\,(\phi_n \cdot f_n(X_1, \ldots, X_n) \neq (X_1, \ldots X_n)).$$

Shannon's **source coding theorem** is as follows.

Theorem 3.6. *Let H be the entropy of the source and $R > H$. There exists a sequence of $(2^{nR_n}, n)$ block codes with error probability $P_e^{(n)}$ such that $P_e^{(n)} \to 0$ and $R_n \to R$.*

More precisely, this is only the positive part of Shannon's theorem telling that any rate $\geq H + \varepsilon$ is achievable under an arbitrary small bound on the probability of error. (The negative part tells that rates $< H$ are not achievable under the same constraint.)

Before we enter the proof let me give an outline of the method of types. Let $\mathbf{x} \in \mathcal{X}^n$. The **type** of $\mathbf{x} = (x_1, x_2, \ldots, x_n) \in \mathcal{X}^n$ is a probability mass function on $\mathcal{X}$. The mass of $x \in \mathcal{X}$ is the relative frequency of x in the sequence $(x_1, x_2, \ldots, x_n)$:

$$P_{\mathbf{x}}(x) := \frac{1}{n} \#\{1 \leq i \leq n : x_i = x\}.$$

The type of a sequence is another name for the empirical distribution. Let $\mathscr{P}_n$ denote the set of all types $P_{\mathbf{x}}$ when $\mathbf{x} \in \mathcal{X}^n$. The elements of $\mathscr{P}_n$ are called n-types. The number of possible n-types is

$$\#(\mathscr{P}_n) = \binom{n + \#(\mathcal{X}) - 1}{\#(\mathcal{X}) - 1} \leq (n+1)^{\#(\mathcal{X})}.$$

The upper estimate is useful in estimations.

For $P \in \mathscr{P}_n$ the **type class** of P is defined as the set of all sequences of type P:

$$\mathrm{Type}(P) := \{\mathbf{x} \in \mathcal{X}^n : P_{\mathbf{x}} = P\}.$$

The cardinality of a type class $\mathrm{Type}(P)$ is a multinomial coefficient

$$\frac{n!}{\prod_x (n P_{\mathbf{x}}(x))!} \qquad (P_{\mathbf{x}} \in P(T))$$

but the following exponential bounds are good enough:

$$\frac{1}{(n+1)^{\#(\mathcal{X})}} 2^{nH(P)} \leq \#(\mathrm{Type}(P)) \leq 2^{nH(P)}. \qquad (3.12)$$

(A proof could be based on Stirling's formula on factorial functions, see [26] p. 282 or [30] p. 430 for other proofs.)

Example 3.1. In this example an application of the method of types is shown.

Let p be a probability measure on $\mathscr{X}$ and set $d := \#(\mathscr{X})$. For each $n \in \mathbb{N}$ and $\delta > 0$ $\Delta(p;n,\delta)$ is the set of all sequences $\mathbf{x} \in \mathscr{X}^n$ such that $|P_{\mathbf{x}}(x) - p(x)| < \delta$ for all $x \in \mathscr{X}$. One can say that $\Delta(p;n,\delta)$ is the set of all δ-typical sequences (with respect to the measure p).

Let $\mu_{n,\delta}$ be the maximizer of the Shannon entropy on the set of all types $P_{\mathbf{x}}$, $\mathbf{x} \in \mathscr{X}^n$, such that $|P_{\mathbf{x}}(x) - p(x)| < \delta$ for every $x \in \mathscr{X}$. We can use the cardinality of the type class corresponding to $\mu_{n,\delta}$ to estimate the cardinality of $\Delta(p; n, \delta)$:

$$(n+1)^{-d} 2^{nH(\mu_{n,\delta})} \leq \#(\Delta(p;n,\delta)) \leq 2^{nH(\mu_{n,\delta})}(n+1)^d,$$

see (3.12). (Indeed, the lower estimate is from the size of type class of $\mu_{n,\delta}$ and the upper estimate is the bound for the number of type classes multiplied with the bound for the previously mentioned type class.) It follows that

$$\lim_{n\to\infty} \frac{1}{n} \log \#\Delta(p;n,\delta) = \sup \left\{ S(q) : |q(x) - p(x)| < \delta \text{ for every } x \in \mathscr{X} \right\},$$

moreover

$$S(p) = \lim_{\delta \to +0} \lim_{n\to\infty} \frac{1}{n} \log \#\Delta(p;n,\delta). \qquad (3.13)$$

The Shannon entropy is obtained from the size of the typical sequences. □

Assume that a probability measure Q on $\mathscr{X}$ is the common distribution of the random variables $X_1, X_2, \ldots, X_n$ and let Q^n be the product measure on $\mathscr{X}^n$, that is, the joint distribution of $X_1, X_2, \ldots, X_n$. The probability of a sequence $\mathbf{x} \in \mathscr{X}^n$ depends only on the type $P_{\mathbf{x}}$ of $\mathbf{x}$. A straight calculation gives that

$$Q^n(\{\mathbf{x}\}) = \prod_x Q(x)^{nP_{\mathbf{x}}(x)} = 2^{-nH(P_{\mathbf{x}}) - nD(P_{\mathbf{x}} \| Q)}.$$

The probability of a type class has exponential bounds:

$$\frac{1}{(n+1)^{\#(\mathscr{X})}} 2^{-nD(P\|Q)} \leq Q^n(\text{Type}(P)) = Q^n(\{\mathbf{x}\}) \times \#(\text{Type}(P)) \leq 2^{-nD(P\|Q)}$$

for $P \in \mathscr{P}_n$.

Proof of Theorem 3.6: Let $\{Q(x) : x \in \mathscr{X}\}$ be the probability distribution of the given source and assume that $R > H(Q)$. Following the idea of Csiszár and Körner [29], set

$$R_n := R - \#(\mathscr{X}) \frac{\log(n+1)}{n}$$

and

$$A_n := \{\mathbf{x} \in \mathscr{X}^n : H(P_{\mathbf{x}}) \leq R_n\}.$$

Then

$$\#(A_n) = \sum \#(T(P)) \le \sum 2^{nH(P)} \le \sum 2^{nR_n}$$
$$\le (n+1)^{\#(\mathcal{X})} 2^{nR_n} = 2^{nR},$$

where all summations are over the set $\{P \in \mathscr{P}_n : H(P) \le R_n\}$.

We can easily define an encoding and a decoding such that elements of A_n are encoded correctly and the other sequences give an error. (Elements of A_n are as just used codewords.) Then the probability of error is

$$P_e^{(n)} = 1 - \text{Prob}(A_n) = \sum Q^n(T(P)),$$

where the summation is over all $P \in \mathscr{P}_n$ such that $H(P) > R_n$. Estimating the sum by the largest term, we can obtain

$$P_e^{(n)} \le (n+1)^{\#(\mathcal{X})} 2^{-n \min S(P\|Q)}, \tag{3.14}$$

where min is over all $P \in \mathscr{P}_n$ such that $H(P) > R_n$. When N is large enough then $R_N > H(Q) + \delta$ and $Q \notin \{P : H(P) \ge R_N\}$. We need a lower bound which does not depend on $n > N$:

$$\min\{D(P\|Q) : P \in \mathscr{P}_n, H(P) \ge R_n\} \ge \min\{D(P\|Q) : H(P) \ge H(Q) + \delta\} > 0.$$

The minimum in the exponent is strictly positive and we can conclude that the probability of error converges to 0 exponentially fast as $n \to \infty$.   □

The interesting feature of the block code constructed in the proof of the theorem is the fact that the distribution Q of the source does not appear, only its entropy $H(Q)$ should be known to construct the **universal encoding scheme**.

3.3 von Neumann Entropy

In the traditional approach to quantum mechanics, a physical system is described in a Hilbert space: Observables correspond to self-adjoint operators, and statistical operators are associated with the states. von Neumann associated an entropy quantity to a statistical operator in 1927 [77] and the discussion was extended in his book [78]. His argument was a gedanken experiment on the grounds of phenomenological thermodynamics which is not repeated here, only his conclusion. Assume that the density ρ is the mixture of orthogonal densities ρ_1 and ρ_2, $\rho = p\rho_1 + (1-p)\rho_2$. Then

$$pS(\rho_1) + (1-p)S(\rho_2) = S(\rho) + \kappa p \log p + \kappa(1-p)\log(1-p), \tag{3.15}$$

where S is a certain thermodynamical entropy quantity, relative to the fixed temperature and molecule density. (Remember that the orthogonality of states has a particular meaning in quantum mechanics, see Example 2.4 in Chap. 2.) From the two-component mixture, we can easily move to an arbitrary density matrix $\rho = \sum_i \lambda_i |\varphi_i\rangle\langle\varphi_i|$ and we have

$$S(\rho) = \sum_i \lambda_i S(|\varphi_i\rangle\langle\varphi_i|) - \kappa \sum_i \lambda_i \log \lambda_i. \tag{3.16}$$

This formula reduces the determination of the (thermodynamical) entropy of a mixed state to that of pure states. The so-called **Schatten decomposition**, $\sum_i \lambda_i |\varphi_i\rangle \langle\varphi_i|$, of a statistical operator is not unique although $\langle \varphi_i, \varphi_j \rangle = 0$ is assumed for $i \neq j$. When λ_i is an eigenvalue with multiplicity, then the corresponding eigenvectors can be chosen in many ways. If we expect the entropy $S(\rho)$ to be independent of the Schatten decomposition, then we are led to the conclusion that $S(|\varphi\rangle\langle\varphi|)$ must be independent of the state vector $|\varphi\rangle$. This argument assumes that there are no super-selection sectors; that is, any vector of the Hilbert space can be a state vector. (Von Neumann's argument was somewhat different, see the original paper [77] or [97].) If the entropy of pure states is defined to be 0 as a kind of normalization, then we have the **von Neumann entropy** formula:

$$S(\rho) = -\kappa \sum_i \lambda_i \log \lambda_i = \kappa \mathrm{Tr}\, \eta(\rho) \tag{3.17}$$

if λ_i are the eigenvalues of ρ and $\eta(t) = -t \log t$. For the sake of simplicity the multiplicative constant κ will mostly be omitted.

After von Neumann, it was Shannon who initiated the interpretation of the quantity $-\sum_i p_i \log p_i$ as "uncertainty measure" or "information measure." Von Neumann himself never made any connection between his quantum mechanical entropy and information. Although von Neumann's entropy formula appeared in 1927, there was not much activity concerning it for several decades.

It is worthwhile to note that if $S(\rho)$ is interpreted as the uncertainty carried by the statistical operator ρ, then (3.15) seems to be natural,

$$S(p\rho_1 + (1-p)\rho_2) = pS(\rho_1) + (1-p)S(\rho_2) + H(p, 1-p), \tag{3.18}$$

holds for an orthogonal mixture and Shannon's classical information measure is involved. The **mixing property** (3.18) essentially determines the von Neumann entropy and tells us that the relation of orthogonal quantum states is classical. A detailed axiomatic characterization of the von Neumann entropy is Theorem 2.1 in [83].

Theorem 3.7. *Let ρ_1 and ρ_2 be density matrices and $0 < p < 1$. The following inequalities hold:*

$$pS(\rho_1) + (1-p)S(\rho_2) \leq S(p\rho_1 + (1-p)\rho_2)$$
$$\leq pS(\rho_1) + (1-p)S(\rho_2) + H(p, 1-p).$$

Proof. The first inequality is an immediate consequence of the concavity of the function $\eta(t) = -t\log t$ (see (11.20)). In order to obtain the second inequality we can use the formula

$$\text{Tr}\,A\big(\log(A+B) - \log A\big)$$
$$= \int_0^\infty \text{Tr}\,A(A+t)^{-1}B(A+B+t)^{-1}dt \geq 0 \qquad (A,B \geq 0)$$

and infer

$$\text{Tr}\,p\rho_1 \log(p\rho_1 + (1-p)\rho_2) \geq \text{Tr}\,p\rho_1 \log p\rho_1$$

and

$$\text{Tr}\,(1-p)\rho_2 \log(p\rho_1 + (1-p)\rho_2) \geq \text{Tr}\,(1-p)\rho_2 \log(1-p)\rho_2.$$

Adding the latter two inequalities, we can obtain the second inequality of the theorem.   □

The von Neumann entropy is the trace of a continuous function of the density matrix, hence it is an obviously continuous functional on the states. However, a more precise estimate for the continuity will be required in approximations.

Theorem 3.8. *Let ρ_1 and ρ_2 be densities on a d-dimensional Hilbert space and let $p := \|\rho_1 - \rho_2\|_1/2$. Then*

$$|S(\rho_1) - S(\rho_2)| \leq p \, \log(d-1) + H(p, 1-p)$$

holds.

Proof. Let $\lambda_1 \geq \lambda_2 \geq \ldots \geq \lambda_n$ and $\mu_1 \geq \mu_2 \geq \ldots \geq \mu_n$ be the eigenvalues of ρ_1 and ρ_2, respectively. Then $S(\rho_1) = H(\lambda_1, \lambda_2, \ldots, \lambda_n)$, $S(\rho_2) = H(\mu_1, \mu_2, \ldots, \mu_n)$ and Lemma 11.1 tells us that

$$\|\rho_1 - \rho_2\|_1 \geq \|(\lambda_1, \lambda_2, \ldots, \lambda_n) - (\mu_1, \mu_2, \ldots, \mu_n)\|_1,$$

therefore it is enough to prove the theorem for probability distributions, or for random variables:

$$H(\lambda_1, \lambda_2, \ldots, \lambda_n) - H(\mu_1, \mu_2, \ldots, \mu_n) \leq p \, \log(d-1) + H(p, 1-p),$$

where p is half of the L^1 distance of the distributions $(\lambda_1, \lambda_2, \ldots, \lambda_n)$ and $(\mu_1, \mu_2, \ldots, \mu_n)$. From the Fano's theorem we have

$$H(X) - H(Y) \leq H(X|Y) \leq q \, \log(d-1) + H(q, 1-q),$$

where X and Y are random variables with distributions $(\lambda_1, \lambda_2, \ldots, \lambda_n)$ and $(\mu_1, \mu_2, \ldots, \mu_n)$; moreover $q = \text{Prob}(X \neq Y)$ (see Theorem 3.3). It isknown in probability theory that the random variables X and Y can be chosen such that $q = p$, this is called "the coupling inequality" (see [74]). This proof is due to Imre **Csiszár**.

□

Note that on an infinite-dimensional Hilbert space the von Neumann entropy is not continuous (but it is restricted to a set $\{\rho : S(\rho) \leq c\}$). Note that an equally useful estimate follows from Theorem 5.1.

Most properties of the von Neumann entropy will be deduced from the behavior of the quantum relative entropy.

3.4 Quantum Relative Entropy

The relative entropy, or I-divergence of the probability distributions $p_1(x)$ and $p_2(x)$, is defined as

$$D(p_1 \| p_2) = \int_{-\infty}^{\infty} p_1(x) \log \frac{p_1(x)}{p_2(x)} \, dx. \tag{3.19}$$

The quantum relative entropy was introduced first in the setting of von Neumann algebras by Umegaki [115] in 1962; it was used in mathematical physics by Lindblad [73] and the definition was extended by Araki to arbitrary von Neumann algebras [7]. Relative entropy showed up in quantum ergodic and information theory not earlier than in the 1980s.

Assume that ρ_1 and ρ_2 are density matrices on a Hilbert space $\mathscr{H}$, then

$$S(\rho_1 \| \rho_2) = \begin{cases} \mathrm{Tr}\,\rho_1(\log \rho_1 - \log \rho_2) & \text{if } \mathrm{supp}\,\rho_1 \leq \mathrm{supp}\,\rho_2 \\ +\infty & \text{otherwise.} \end{cases} \tag{3.20}$$

The **relative entropy** expresses statistical distinguishability and therefore it decreases under stochastic mappings. Note that it is not a symmetric function of the two arguments. To provide some motivation to study this quantity, I present an example (which is nothing else but the quantum Stein lemma).

Example 3.2. In the **hypothesis testing** problem we have to decide between the states ρ_0 and ρ_1. The first state is the **null hypothesis** and the second one is the **alternative hypothesis**. The decision is performed by a two-valued measurement $\{P, I - P\}$, where the projection P corresponds to the acceptance of ρ_0 and $I - P$ corresponds to the acceptance of ρ_1. P is called **test**. $\alpha := \mathrm{Tr}\,\rho_0(I - P)$ is the **error of the first kind**. This is the probability that the null hypothesis is true but we have to decide the alternative hypothesis. $\beta := \mathrm{Tr}\,\rho_1 P$ is the **error of the second kind**, which is the probability that the alternative hypothesis is true but we choose the null hypothesis.

The problem is to decide which hypothesis is true in an asymptotic situation, where the n-fold product states $\rho_0^{(n)}$ and $\rho_1^{(n)}$ are at our disposal. The decision is performed by a test P_n. The errors of the first and second kind depend on n.

Set

$$\beta^*(n,\varepsilon) := \inf\{\mathrm{Tr}\,\rho_1^{(n)}P_n : \mathrm{Tr}\,\rho_0^{(n)}(I - P_n) \leq \varepsilon\},$$

which is the infimum of the error of the second kind when the error of the first kind is at most ε. In mathematical statistics ε is usually prescribed and the sample size n is chosen to make the error of the first kind to be small. The minimal error of second kind converges to 0 exponentially fast:

$$\lim_{n\to\infty} \frac{1}{n}\log\beta^*(n,\varepsilon) = -S(\rho_0\|\rho_1) \qquad (3.21)$$

is known as **quantum Stein lemma**, (see Theorem 8.1).

In this hypothesis testing problem the null hypothesis and the alternative hypothesis play different roles, which corresponds to the fact that the relative entropy is not symmetric in its two variables.   □

Example 3.3. Let ρ be an $n \times n$ density matrix. I/n is the density of the tracial state. Then

$$S(\rho\|I/n) = \log n - S(\rho).$$

Therefore up to some constants, the relative entropy is the extension of the von Neumann entropy. If the background state is uniform, then the relative entropy reduces to the von Neumann entropy.   □

Theorem 3.9. *Let ρ_1 and ρ_2 be density matrices in $B(\mathscr{H})$ and let $\mathscr{E} : B(\mathscr{H}) \to B(\mathscr{K})$ be a state transformation. Then the monotonicity*

$$S(\rho_1\|\rho_2) \geq S(\mathscr{E}(\rho_1)\|\mathscr{E}(\rho_2))$$

holds.

The presented proof, which is based on the relative modular operator method, follows [88].

Let ρ_1 and ρ_2 be density matrices acting on the Hilbert space $\mathscr{H}$ and assume that they are invertible. The set $B(\mathscr{H})$ of bounded operators acting on $\mathscr{H}$ becomes a Hilbert space when the Hilbert–Schmidt inner product

$$\langle A, B \rangle := \mathrm{Tr}\,A^*B$$

is regarded. On the Hilbert space $B(\mathscr{H})$, one can define an operator $\Delta(\rho_2/\rho_1) \equiv \Delta$ as

$$\Delta a = \rho_2 a \rho_1^{-1} \qquad (a \in B(\mathscr{H})). \qquad (3.22)$$

(If ρ_1 is not invertible, then ρ_1^{-1} is a generalized inverse defined on the range of ρ_1, $\rho_1\rho_1^{-1} = \rho_1^{-1}\rho_1 = \mathrm{supp}\,\rho_1$.) This is the so-called **relative modular operator** and it is the product of two commuting positive operators: $\Delta = LR$, where

$$La = \rho_2 a \quad \text{and} \quad Ra = a\rho_1^{-1} \quad (a \in B(\mathcal{H})).$$

Since $\log \Delta = \log L + \log R$, we have

$$\text{Tr}\,\rho_1(\log \rho_1 - \log \rho_2) = -\langle \rho_1^{1/2}, (\log \Delta)\rho_1^{1/2}\rangle.$$

The relative entropy $S(\rho_1 \| \rho_2)$ is expressed by the quadratic form of the logarithm of the relative modular operator. This is the fundamental formula that we use (and actually this is nothing else but Araki's definition of the relative entropy in a general von Neumann algebra [7]). Replacing $-\log$ by a function $f : \mathbb{R}^+ \to \mathbb{R}$, the generalization

$$S_f(\rho_1 \| \rho_2) = \langle \rho_1^{1/2}, f(\Delta(\rho_2/\rho_1))\rho_1^{1/2}\rangle \tag{3.23}$$

is introduced and is called **quasi-entropy**. This generalization is the quantum mechanical counterpart of the f-entropy introduced by Csiszár for probability distributions [27]. The monotonicity holds for the quasi-entropies under some condition on the parameter function f.

Theorem 3.10. *Let ρ_1 and ρ_2 be density matrices in $B(\mathcal{H})$ and let $\mathcal{E} : B(\mathcal{H}) \to B(\mathcal{K})$ be a state transformation. For an operator monotone decreasing continuous function $f : \mathbb{R}^+ \to \mathbb{R}$, the monotonicity*

$$S_f(\rho_1 \| \rho_2) \geq S_f(\mathcal{E}(\rho_1) \| \mathcal{E}(\rho_2))$$

holds.

Proof. For the sake of simplicity we should assume that all the densities have only non-zero eigenvalues. The general case can be covered by an approximation argument. Due to the simple transformation formula $S_{f+c} = S_f + c$ for a real constant c, we may assume that $f(0) = 0$.

Set the relative modular operators Δ and Δ_0 on the spaces $B(\mathcal{H})$ and $B(\mathcal{K})$, respectively, as follows.

$$\Delta a = \rho_2 a \rho_1^{-1} \quad (a \in B(\mathcal{H})) \quad \text{and} \quad \Delta_0 x = \mathcal{E}(\rho_2)x\mathcal{E}(\rho_1)^{-1} \quad (x \in B(\mathcal{K})).$$

Note that both spaces become a Hilbert space with the Hilbert–Schmidt inner product. $\mathcal{E}^* : B(\mathcal{K}) \to B(\mathcal{H})$ stands for the adjoint of $\mathcal{E}$.

The operator

$$Vx\mathcal{E}(\rho_1)^{1/2} = \mathcal{E}^*(x)\rho_1^{1/2} \quad (x \in B(\mathcal{K})) \tag{3.24}$$

is a contraction:

$$\|\mathcal{E}^*(x)\rho_1^{1/2}\|^2 = \text{Tr}\,\rho_1\mathcal{E}^*(x^*)\mathcal{E}^*(x) \leq \text{Tr}\,\rho_1\mathcal{E}^*(x^*x) = \text{Tr}\,\mathcal{E}(\rho_1)x^*x = \|x\mathcal{E}(\rho_1)^{1/2}\|^2$$

since the Schwarz inequality is applicable to $\mathcal{E}^*$. A similar simple computation gives that

$$V^*\Delta V \leq \Delta_0. \tag{3.25}$$

Since f is operator monotone decreasing, we have $f(\Delta_0) \leq f(V^*\Delta V)$. Recall that f is operator convex, therefore $f(V^*\Delta V) \leq V^* f(\Delta) V$ and we shall conclude

$$f(\Delta_0) \leq V^* f(\Delta) V. \tag{3.26}$$

(See the Appendix about operator monotone functions and related inequalities). Since $V \mathscr{E}(\rho_1)^{1/2} = \rho_1^{1/2}$, this implies

$$\langle \mathscr{E}(\rho_1)^{1/2}, f(\Delta_0)\mathscr{E}(\rho_1)^{1/2} \rangle \leq \langle \rho_1^{1/2}, f(\Delta)\rho_1^{1/2} \rangle,$$

which is our statement.   □

Now let us return to the proof of Theorem 3.9. In order to deduce the monotonicity theorem for the ordinary relative entropy, we can apply Theorem 3.10 to the function $f_\varepsilon(t) = -\log \frac{t+\varepsilon}{\varepsilon}$ for $\varepsilon \to 0$.   □

Example 3.4. Assume that an inequality

$$S_f(\rho_1 \| \rho_2) \geq G(\|\rho_1 - \rho_2\|_1)$$

holds classically, that is, for commuting ρ_1 and ρ_2. Then the inequality holds generally in the quantum case.

For example.

$$2D(\mu_1 \| \mu_2) \geq (\|\mu_1 - \mu_2\|_1)^2 \tag{3.27}$$

is the **Pinsker–Csiszár inequality**, which extends to the quantum case.

Consider the commutative subalgebra generated by ρ_1 and ρ_2. The reduction of ρ_1 and ρ_2 can be viewed as probability distributions μ_1 and μ_2 and $\|\rho_1 - \rho_2\|_1 = \|\mu_1 - \mu_2\|_1$. From the monotonicity of the quasi-entropy S_f, we have

$$S_f(\rho_1 \| \rho_2) \geq S_f(\mu_1 \| \mu_2) \geq G(\|\mu_1 - \mu_2\|_1) = G(\|\rho_1 - \rho_2\|_1).$$

In particular,

$$2S(\rho_1 \| \rho_2) \geq (\|\rho_1 - \rho_2\|_1)^2. \tag{3.28}$$

A different estimate is (11.22).   □

Example 3.5. It follows from the monotonicity (or from the definition directly) that the relative entropy is invariant under unitary conjugation:

$$S(\rho \| \omega) = S(U\rho U^* \| U\omega U^*)$$

for every unitary U.

Consider the densities

$$\rho := \tfrac{1}{2}(I + u \cdot \sigma) \quad \text{and} \quad \omega := \tfrac{1}{2}(I + v \cdot \sigma)$$

of a qubit, where u and v are vectors in the Bloch ball. Due to the rotation invariance (coming from the unitary invariance), the relative entropy $S(\rho \| \omega)$ depends on $\|u\|, \|v\|$ and on the angle of u and v. By rotation we can assume that both u and v

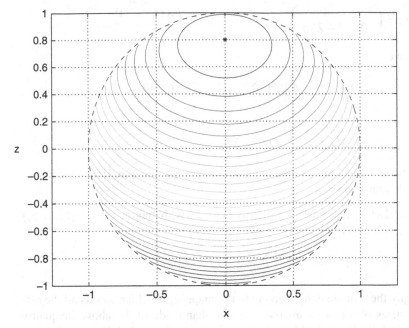

Fig. 3.1 The levels of the relative entropy of $\rho = \frac{1}{2}(I + (x, 0, z) \cdot \sigma)$ with respect to the density $\omega := \frac{1}{2}(I + (0, 0, 0.8) \cdot \sigma)$

are in the x–y plane, $v = (x_v, 0, 0)$ and $u = (x_u, y_u, 0)$. In Fig. 3.1 one can see the level of the relative entropy with respect to ω when $v = (0.8, 0, 0)$. $\square$

The monotonicity is a very strong property. The monotonicity of the quasi-entropy implies some other general properties.

Theorem 3.11. *For an operator monotone decreasing function* $f : \mathbb{R}^+ \to \mathbb{R}$, *the quasi-entropy* $S_f(\rho_1, \rho_2)$ *is jointly convex,*

$$S_f\big(\lambda\rho_1 + (1-\lambda)\omega_1 \| \lambda\rho_2 + (1-\lambda)\omega_2\big) \leq \lambda S_f(\rho_1\|\rho_2) + (1-\lambda)S_f(\omega_1\|\omega_2)$$

and

$$S_f(\rho_1\|\rho_2) \geq f(1).$$

When f is not affine, then the equality holds here if and only if $\rho_1 = \rho_2$.

Concerning the full proof, refer to the original paper [88] or the monograph [83]. Since the **joint convexity** of the relative entropy is rather important, I will show how to deduce it from the monotonicity.

Let ρ_1, ρ_2, ω_1 and ω_2 be density matrices on a Hilbert space $\mathscr{H}$. Consider the following density matrices on $\mathscr{H} \otimes \mathbb{C}^2$:

$$\rho := \begin{pmatrix} \lambda \rho_1 & 0 \\ 0 & (1-\lambda)\rho_2 \end{pmatrix} \quad \text{and} \quad \omega := \begin{pmatrix} \lambda \omega_1 & 0 \\ 0 & (1-\lambda)\omega_2 \end{pmatrix}.$$

Then we have

$$S(\rho \| \omega) = \lambda S(\rho_1 \| \rho_2) + (1-\lambda)S(\omega_1 \| \omega_2)$$

and

$$S(\lambda \rho_1 + (1-\lambda)\omega_1 \| \lambda \rho_2 + (1-\lambda)\omega_2)$$

is exactly the relative entropy of the reduced densities of ρ and ω. Hence the monotonicity yields the convexity.

Example 3.6. Let ρ_1 and ρ_2 be $n \times n$ density matrices with diagonal $(p_1, p_2, \ldots, p_n)$ and $(q_1, q_2, \ldots, q_n)$, respectively. Then

$$S(\rho_1 \| \rho_2) \geq \sum_i p_i (\log p_i - \log q_i).$$

We can apply the monotonicity theorem to the mapping $\mathscr{E}$ which annuls all the off-diagonal entries of a density matrix. The right-hand side of the above inequality is the classical Kullback–Leibler relative entropy of the probability distributions $(p_1, p_2, \ldots, p_n)$ and $(q_1, q_2, \ldots, q_n)$.

Choosing $\rho_2 = I/n$, we have the entropy inequality

$$S(\rho_1) \leq H(p_1, p_2, \ldots, p_n), \tag{3.29}$$

the von Neumann entropy is majorized by the Shannon entropy of the diagonal. □

This example might be generalized to a measurement.

Example 3.7. Let ρ_1 and ρ_2 be statistical operators acting on the Hilbert space $\mathscr{H}$. Suppose that a measurement is given by the positive operators $F_x \in B(\mathscr{H})$ such that $\sum_x F_x = I_{\mathscr{H}}$. ρ_1 and ρ_2 induce the a posteriori distributions $p(x) = \mathrm{Tr}\rho_1 F_x$ and $q(x) = \mathrm{Tr}\rho_2 F_x$. Then

$$S(\rho_1 \| \rho_2) \geq D(p(x) \| q(x)).$$

Measurement is a coarse-graining, and a loss of information takes place. □

Theorem 3.12. *Let ρ_{12} and ω_{12} be density matrices acting on the Hilbert space $\mathscr{H}_1 \otimes \mathscr{H}_2$. Assume that $\rho_{12} = \rho_1 \otimes \rho_2$ is a product and let ω_1 be the reduced state of ω_{12}. Then*

$$S(\omega_{12} \| \rho_{12}) = S(\omega_1 \| \rho_1) + S(\omega_{12} \| \omega_1 \otimes \rho_2).$$

The theorem tells us that the difference $S(\omega_{12} \| \rho_{12})$ of ρ_{12} and ω_{12} comes from two reasons. The difference of the two states reduced to the first system and the conditioned difference with respect to the first system. (ω and ω' have the same reduced density on $\mathscr{H}_1$.) The proof is easy and follows from the definitions. The identity is called **conditional expectation property**, since the existence of the conditional expectation preserving ρ_{12} is assumed. A possible axiomatization of the relative entropy is based on the conditional expectation property (see Chap. 2 in [83]).

Example 3.8. The invertible density matrices form a manifold which has two natural parameterizations, affine and exponential. (The latter appears when the density is a Gibbs state corresponding to a self-adjoint Hamiltonian.)

Let $\rho \equiv e^H$, ω_1 and ω_2 be three invertible densities. The *e*-**geodesic** connecting ρ and ω_2 is the curve

$$\gamma_e(t) = \frac{\exp(H+tA)}{\mathrm{Tr}\exp(H+tA)} \qquad (t \in [0,1]),$$

where $A = \log\omega_2 - \log\rho$. Then $\gamma_e(0) = \rho$ and $\gamma_e(1) = \omega_2$. The *m*-**geodesic** connecting ρ and ω_1 is the curve

$$\gamma_m(t) = (1-t)\rho + t\omega_1 = \rho + t(\omega_1 - \rho) \quad (t \in [0,1]),$$

$\gamma_m(0) = \rho$ and $\gamma_e(1) = \omega_1$. The *e*-geodesic is the natural path in the exponential parameterization and the *m*-geodesic corresponds to the mixture or affine parameterization. If we want to compute the tangent vector of curve, we should choose a parameterization. In the exponential parameterization

$$\frac{\partial}{\partial t}\log\gamma(t)\Big|_{t=0}$$

is the tangent vector at $t = 0$ for a curve γ. In case of the exponential geodesic we can get A trivially. For the mixture geodesic,

$$\frac{\partial}{\partial t}\log(\rho + t(\rho - \omega_1))\Big|_{t=0} = \frac{\partial}{\partial t}\int_0^\infty (1+s)^{-1} - (s+\rho+t\rho-t\omega_1)^{-1}\,ds$$

$$= \int_0^\infty (\rho+s)^{-1}(\omega_1 - \rho)(\rho+s)^{-1}\,ds.$$

Assume that the *e*-geodesic connecting ρ and ω_2 is orthogonal to the *m*-geodesic connecting ρ and ω_1 with respect to the inner product

$$\langle B,C\rangle_\rho := \int_0^\infty \mathrm{Tr}\,(\rho+s)^{-1}B^*(\rho+s)^{-1}C\,ds. \qquad (3.30)$$

A plain computation yields

$$S(\omega_1,\rho) + S(\rho,\omega_2) - S(\omega_1,\omega_2) = \mathrm{Tr}A(\omega_1 - \rho) = \langle A, \omega_1 - \rho\rangle_{HS}.$$

For the superoperator

$$\mathbf{T}_\rho : X \mapsto \int_0^\infty (\rho + s)^{-1} X (\rho + s)^{-1} \, ds \tag{3.31}$$

we have

$$\langle X, Y \rangle_{HS} = \langle \mathbf{T}_\rho(X), Y \rangle_\rho . \tag{3.32}$$

Therefore,

$$\langle A, \omega_1 - \rho \rangle_{HS} = \langle T_\rho(A), \omega_1 - \rho \rangle_\rho = \langle \dot{\gamma}_e(0), \dot{\gamma}_m(0) \rangle_\rho$$

and we can conclude that

$$S(\omega_1, \rho) + S(\rho, \omega_2) = S(\omega_1, \omega_2). \tag{3.33}$$

This relation is sometimes called **Pythagorean theorem** (Fig. 3.2). (The relative entropy plays the role of the square of the Euclidean distance.)

In the light of the Pythagorean theorem, Theorem 3.12 has a very nice geometric interpretation. Let S be the convex set of all densities on $\mathscr{H}_1 \otimes \mathscr{H}_2$ which have the form $D \otimes \rho_2$. The density ω_{12} has a best approximation from S and it is $\omega_1 \otimes \rho_2$. The three densities ω_{12}, $\omega_1 \otimes \rho_2$ and $\rho_1 \otimes \rho_2$ form a triangle. The relation

$$S(\omega_{12} \| \rho_{12}) = S(\omega_1 \otimes \rho_2 \| \rho_1 \otimes \rho_2) + S(\omega_{12} \| \omega_1 \otimes \rho_2)$$

is equivalent to the statement of the theorem and shows that the triangle is rectangular.   □

Example 3.9. We can use the notation

$$\lambda_n := 1/n \quad \text{and} \quad \mu_n = 1 - \lambda_n \quad (n \in \mathbb{N}).$$

Let ρ and σ be invertible density matrices. Then

$$S(\rho \| \mu_n \rho + \lambda_n \sigma) \to 0$$

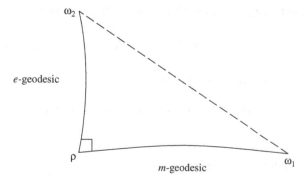

Fig. 3.2 The Pythagorean theorem for relative entropy: $S(\omega_1, \rho) + S(\rho, \omega_2) = S(\omega_1, \omega_2)$

as $n \to \infty$, since $\mu_n \rho + \lambda_n \sigma \to \rho$. We shall prove that the convergence is fast, that is,

$$nS(\rho \| \mu_n \rho + \lambda_n \sigma) \to 0.$$

As per the definition of the relative entropy, we have

$$nS(\rho \| \mu_n \rho + \lambda_n \sigma) = n\mathrm{Tr}\,\rho \left(\log \rho - \log(\mu_n \rho + \lambda_n \sigma) \right).$$

Using the formula

$$\log x = \int_0^\infty \frac{1}{1+t} - \frac{1}{t+x}\, dt$$

we can transform the right-hand side:

$$nS(\rho \| \mu_n \rho + \lambda_n \sigma) = n \int_0^\infty \mathrm{Tr}\,\rho \left(-(t+\rho)^{-1} + (t+\mu_n \rho + \lambda_n \sigma)^{-1} \right) dt$$
$$= \int_0^\infty \mathrm{Tr}\,\rho (t+\rho)^{-1}(\rho - \sigma)\left(t + \mu_n \rho + \lambda_n \sigma \right)^{-1} dt.$$

The integrand is majorized by an integrable function $C(t+\varepsilon)^{-2}$ and has the limit

$$\mathrm{Tr}\,\rho (t+\rho)^{-2}(\rho - \sigma)$$

as $n \to \infty$. The integral of this quantity can be computed and we can get $\mathrm{Tr}\,(\rho - \sigma)$, which is 0.   □

3.5 Rényi Entropy

The **Rényi entropy** of order $\alpha \neq 1$ of the probability distribution $(p_1, p_2, \ldots, p_n)$ is defined by

$$H_\alpha(p_1, p_2, \ldots, p_n) = \frac{1}{1-\alpha} \log \sum_{k=1}^n p_k^\alpha. \tag{3.34}$$

(In this section log denotes the logarithm of base 2.) The limit $\alpha \to 1$ recovers the Shannon entropy. If $(p_1, p_2, \ldots, p_n)$ is the distribution of a random variable X, then instead of (3.34) we may write $H_\alpha(X)$.

It follows from the definition that the Rényi entropies are **additive**: When X and Y are independent random variables, then

$$H_\alpha(X, Y) = H_\alpha(X) + H_\alpha(Y). \tag{3.35}$$

The additivity is crucial in **axiomatization**.

Theorem 3.13. *Assume that for any collection $(p_1, p_2, \ldots, p_n)$ of positive numbers such that $0 < \sum_i p_i \leq 1$ a function $H(p_1, p_2, \ldots, p_n)$ is defined and has the following properties:*

(1) *H is symmetric and continuous.*
(2) *H is additive (3.35).*
(3) *H has the decomposition property:*

$$H(p_1, p_2, \ldots, p_n, q_1, q_2, \ldots, q_n) = g^{-1}(\beta H(p_1, p_2, \ldots, p_n)$$
$$+ (1-\beta)H(q_1, q_2, \ldots, q_n)),$$

where

$$\beta = \frac{\Sigma_i\, p_i}{\Sigma_i\, p_i + \Sigma_i\, q_i}.$$

If $g(x) \equiv 1$, then (up to a constant factor)

$$H(p_1, p_2, \ldots, p_n) = -\frac{\Sigma_i\, p_i \log p_i}{\Sigma_i\, p_i}.$$

If $g(x) = 2^{(\alpha-1)x}$ with $1 \neq \alpha > 0$, then

$$H(p_1, p_2, \ldots, p_n) = H_\alpha(p_1, p_2, \ldots, p_n).$$

$\square$

The Rényi entropies can be defined by optimal coding, similar to the Shannon entropy.

Let X be a random variable with a finite range $\mathscr{X}$. A source code C for X is a mapping from $\mathscr{X}$ to the set of finite-length strings of symbols of a binary alphabet. Let $C(x)$ denote the codeword corresponding to x and let $\ell(x)$ denote the length of $C(x)$. If $p(x)$ is the probability of $x \in \mathscr{X}$, then we can define

$$L_\beta(C) := \frac{1}{\beta} \log \sum_x p(x) 2^{\beta \ell(x)}$$

to be the cost of the coding. When the code is uniquely decodable, the Kraft–MacMillan inequality tells us that

$$\sum_x 2^{-\ell(x)} \leq 1.$$

The Hölder inequality

$$\sum_j x_j y_j \geq \left(\sum_j x_j^p\right)^{1/p} \left(\sum_j y_j^q\right)^{1/q}$$

holds when $p < 1$ and $q = p/(p-1)$. For $\beta > -1$, this gives

$$\sum_x 2^{-\ell(x)} = \sum_x (p_j^{-1/\beta} 2^{-\ell(x)}) p_j^{1/\beta} \geq \left(\sum_x p(x) 2^{\beta \ell(x)}\right)^{-1/\beta} \left(\sum_x p(x)^\alpha\right)^{1/(1-\alpha)},$$

where $\alpha = 1/(1+\beta)$. Since the left-hand side is bounded by 1, we have

$$\left(\sum_x p(x) 2^{\beta \ell(x)}\right)^{1/\beta} \geq \left(\sum_x p(x)^{\alpha}\right)^{1/(1-\alpha)}.$$

After taking the logarithm, we can conclude the first half of the following result of Campbell [23].

Theorem 3.14. *Let $\beta > -1$. For a uniquely decodable code, the inequality*

$$L_\beta(C) \geq H_\alpha(X)$$

holds when $\alpha = 1/(1+\beta)$. Moreover, there exists a uniquely decodable code C such that

$$L_\beta(C) \leq H_\alpha(X) + 1.$$

This theorem actually gives the value of the Rényi entropy only approximately. To have a complete determination, one has to pass to multiple use of the source.

The Rényi entropies are strongly related to the L^p-norm of a probability vector:

$$\|(x_1, x_2, \ldots, x_n)\|_p := \left(\sum_{i=1}^{n} |x_i|^p\right)^{1/p}$$

Since the L^2-norm is the most convenient in several situations, the entropy of order 2

$$H_2(p) = -2\log \|p\|_2$$

is frequently used.

For the Rényi entropies, the interesting value of the parameter is $\alpha > 0$. One can check that

$$\frac{\partial}{\partial \alpha} H_\alpha \leq 0, \quad \frac{\partial}{\partial \alpha}(1-\alpha)H_\alpha \leq 0, \quad \frac{\partial^2}{\partial \alpha^2}(1-\alpha)H_u \geq 0. \tag{3.36}$$

Therefore, H_α is a decreasing function and $(1-\alpha)H_\alpha$ is a convex function of the parameter α.

The **entropy of degree** α is defined as

$$H^\alpha(p_1, p_2, \ldots, p_n) = \frac{1}{2^{1-\alpha} - 1}\left(\sum_{k=1}^{n} p_k^\alpha - 1\right). \tag{3.37}$$

It is easy to transform H^α and H_α into each other but the two quantities have different properties. The limit $\alpha \to 1$ gives back the Shannon entropy again.

H^α is a **symmetric** function of its variables and satisfy the **recursion**

$$\begin{aligned} H^\alpha(p_1, p_2, \ldots, p_n) &= H^\alpha(p_1 + p_2, p_3, \ldots, p_n) \\ &\quad + (p_1 + p_2)^\alpha H^\alpha(p_1/(p_1+p_2), p_2/(p_1+p_2)). \end{aligned}$$

These two properties together with a normalization characterize the entropy of degree α, (see p. 189 in [2]). The usual additivity (3.35) does not hold, but we have the so-called **additivity of degree** α:

$$H^\alpha(X,Y) = H^\alpha(X) + H^\alpha(Y) + (2^{1-\alpha} - 1)H^\alpha(X)H^\alpha(Y), \qquad (3.38)$$

when X and Y are independent. When the value of the parameter α is kept fixed, a constant multiple of H^α can be equally good. In the literature several constant multiples appear. For example,

$$\frac{2^{1-\alpha} - 1}{1 - \alpha} H^\alpha$$

was popularized by Tsallis [112]. For the sake of simplicity, I insist on the original notation H^α.

The quantum analogues of H^α and H_α are as follows.

$$S^\alpha(\rho) = \frac{1}{2^{1-\alpha} - 1}\left(\text{Tr}\rho^\alpha - 1\right). \qquad (3.39)$$

and

$$S_\alpha(\rho) = \frac{1}{1 - \alpha} \log \text{Tr}\rho^\alpha \qquad (3.40)$$

for a density matrix ρ.

The relative α-entropy is defined as

$$S_\alpha(\rho_1\|\rho_2) = \frac{1}{\alpha(1 - \alpha)}\text{Tr}\,(I - \rho_2^\alpha\rho_1^{-\alpha})\rho_1. \qquad (3.41)$$

This is a particular quasi-entropy corresponding to the function

$$f_\alpha(t) = \frac{1}{\alpha(1 - \alpha)}\left(1 - t^\alpha\right),$$

which is operator monotone decreasing for $\alpha \in (-1, 1)$. (For $\alpha = 0$, the limit is taken as $f_0(t) = -\log t$.)

It follows from Theorem 3.10 that the relative α-entropy is monotone under coarse-graining:

$$S_\alpha(\rho_1\|\rho_2) \geq S_\alpha(\mathscr{E}(\rho_1)\|\mathscr{E}(\rho_2)).$$

If follows also from the general properties of quasi-entropies that $S_\alpha(\rho_1\|\rho_2)$ is jointly convex and positive.

3.6 Notes

The von Neumann entropy was introduced by von Neumann in 1927, earlier than Shannon's work about information theory. When Shannon invented his quantity and consulted von Neumann on what to call it, von Neumann replied, *"Call it entropy. It is already in use under that name and besides, it will give you a great edge in debates because nobody knows what entropy is anyway."*

The relative entropy of measures was introduced by Kullback and Leibler in 1951 in connection with sufficiency. "Kullback–Leibler distance" and "information divergence" are other frequently used names [26, 29].

The **quasi-entropies** were introduced by Petz as a quantum counterpart of the f-divergences of Csiszár [87, 88]. Actually, the definition was slightly more general:

$$S_f^K(\rho_1\|\rho_2) = \langle K\rho_1^{1/2}, f(\Delta(\rho_2/\rho_1))K\rho_1^{1/2}\rangle, \qquad (3.42)$$

where K is a fixed matrix and $\Delta(\rho_2/\rho_1)$ is the relative modular operator. For example, for $f(x) = x^t$ we have

$$S_f^K(\rho_1\|\rho_2) = \operatorname{Tr} K^*\rho_2^t K\rho_1^{1-t}. \qquad (3.43)$$

The monotonicity theorem holds in the setting in the form

$$S_f^{\mathscr{E}^*(K)}(\rho_1\|\rho_2) \le S_f^K(\mathscr{E}(\rho_1)\|\mathscr{E}(\rho_2)), \qquad (3.44)$$

where $\mathscr{E}^*$ is the unital adjoint of $\mathscr{E}$ and f is an operator monotone function. Monotonicity implies joint concavity. In particular, the joint concavity of (3.43) is known as **Lieb's concavity** theorem.

With regard to inequality (3.27), it can be noted that Csiszár improved the constant in the inequality due to Pinsker in 1967 (see [27]). The quantum version (3.28) in the von Neumann algebra setting was established in [52].

It was conjectured by Diósi, Feldmann and Kosloff [34] that

$$S(R_n) - (n-1)S(\rho) - S(\sigma) \to S(\sigma\|\rho), \qquad (3.45)$$

where

$$R_n := \frac{1}{n}\left(\sigma\otimes\rho^{\otimes(n-1)} + \rho\otimes\sigma\otimes\rho^{\otimes(n-2)} + \cdots + \rho^{\otimes(n-1)}\otimes\sigma\right)$$

as $n \to \infty$. The proof was given by Csiszár, Hiai and Petz [31].

Details about entropy and relative entropy in the von Neumann algebra setting are in the monograph [83].

The exponential and affine parameterization of the state space are dual. Details of their relation and the Pythagorean theorem are important features of the information geometry (see [5]). Formula (3.30) gives the Kubo–Mori inner product in the exponential parameterization.

For probability distributions, the Rényi entropy was introduced in 1957; however, the work [104] is the standard reference, the axiomatization in Theorem 3.13 is contained here. The entropy of degree α was introduced in 1967 by Havrda and Charvát [48]. The monographs [2] contains the details.

3.7 Exercises

1. Use the **Stirling formula**

$$\sqrt{2\pi n}\left(\frac{n}{e}\right)^n \leq n! \leq \sqrt{2\pi n}\left(\frac{n}{e}\right)^n \exp(1/12n) \tag{3.46}$$

   to show that

$$\lim_{n\to\infty} \frac{1}{n}\log\frac{n!}{(p_1 n)!(p_2 n)!\ldots(p_k n)!} = -\sum_{i=1}^{k} p_i \log p_i \equiv H(p_1, p_2, \ldots, p_k)$$

   for a probability distribution $(p_1, p_2, \ldots, p_k)$.
2. Use the fact that $f(t) = \log t$ is strictly concave to show that

$$\sum_{i=1}^{k} p_i(\log p_i - \log q_i) \geq 0,$$

   if $(p_1, p_2, \ldots, p_k)$ and $(q_1, q_2, \ldots, q_k)$ are the probability distributions.
3. Compute the relative entropy of a probability distribution on $\mathbb{R}$ with respect to a **Gaussian distribution**

$$p_{m,\sigma}(x) := \frac{1}{\sigma\sqrt{2\pi}}\exp\left(-\frac{(x-m)^2}{2\sigma^2}\right).$$

4. Compute the **differential entropy**

$$H(p) := -\int_{-\infty}^{\infty} p(x)\log p(x)\,dx$$

   of a Gaussian distribution. Show that if the mean m and the variance σ are fixed, then the Gaussian distribution maximizes the differential entropy.
5. Show that

$$\log\frac{I + x\cdot\sigma}{2} = \frac{I}{2}\log\frac{1 - \|x\|^2}{4} + (\tanh^{-1}\|x\|)\frac{x}{\|x\|}\cdot\sigma$$

   for $x \in \mathbb{R}^3$ with $\|x\| < 1$. (Hint: Use (2.36).)
6. Use Theorem 3.10 to show that $S_f(\rho_1\|\rho_2) \geq f(1)$ holds for a quasi-entropy S_f and for arbitrary densities ρ_1 and ρ_2.
7. Let $f(x) = (1+x)^{-1}$. Show that

$$S_f(\rho_1\|\rho_2) = \int_0^\infty \mathrm{Tr}\exp(-t\rho_2)\rho_1^2\exp(-t\rho_1)\,dt\,.$$

Simplify the formula for commuting ρ_1 and ρ_2.

8. Prove that $H(X|Y) = H(X)$ if and only if the random variables X and Y are independent.

9. Show that equality holds in (3.29) if and only if the matrix ρ is diagonal.

10. Let $\mathscr{A}$ be a subalgebra of $B(\mathscr{H})$ such that $\mathscr{A}^* = \mathscr{A}$ and $I_{\mathscr{H}} \in \mathscr{A}$ and assume that $\rho \in B(\mathscr{H})$ is a density operator with non-zero eigenvalues. Show that the conditional expectation $E : B(\mathscr{H}) \to \mathscr{A}$ preserving ρ is an orthogonal projection with respect to the inner product $\langle A, B\rangle := \mathrm{Tr}\,\rho A^* B$.

11. Let $\rho \in M_n(\mathbb{C})$ be a density matrix. Show that the ρ-preserving conditional expectation from the full matrix algebra onto the diagonal matrices exists if and only if ρ is diagonal.

12. Prove relation (3.32).

13. Let $\beta > 0$ and H be a self-adjoint matrix. Use the relative entropy to show that the minimum of the functional

$$F(\rho) = \mathrm{Tr}\,\rho H - \frac{1}{\beta}S(\rho)$$

defined on density matrices is reached at

$$\frac{1}{Z}\exp(-\beta H),$$

where the constant Z is for normalization. (This means that the minimum of the free energy is at the Gibbs state.)

Chapter 4
Entanglement

It is a widely accepted statement that entanglement is one of the most striking features of quantum mechanics. If two quantum systems interacted sometime in the past, then it is not possible to assign a single state vector to the subsystems in the future. This is one way to express entanglement that was historically recognized by Einstein, Podolsky and Rosen and by Schrödinger. Later Bell showed that entanglement manifests in non-locality of quantum mechanics. The information theoretic aspect of entanglement was observed by Schrödinger himself: *"Best possible knowledge of a whole does not include best possible knowledge of its parts."* In this formulation entanglement does not sound so striking, an individual component of an entangled system may exhibit more disorder than the whole system.

4.1 Bipartite Systems

The physical setting to see entanglement is the **bipartite system** which corresponds to tensor product in mathematical terms. Let $B(\mathcal{H}_A)$ and $B(\mathcal{H}_B)$ be the algebras of bounded operators acting on the Hilbert spaces $\mathcal{H}_A$ and $\mathcal{H}_B$. The Hilbert space of the composite system $A + B$ is $\mathcal{H}_{AB} := \mathcal{H}_A \otimes \mathcal{H}_B$. The algebra of the operators acting on $\mathcal{H}_{AB}$ is $B(\mathcal{H}_{AB}) = B(\mathcal{H}_A) \otimes B(\mathcal{H}_B)$.

Let us recall how to define ordering in a vector space V. A subset $V_+ \subset V$ is called **positive cone** if $v, w \in V_+$ implies $v + w \in V_+$ and $\lambda v \in V_+$ for any positive real λ. Given the positive cone V_+, $f \leq g$ means that $g - f \in V_+$. In the vector space $B(\mathcal{H})$ the standard positive cone is the set of all positive semidefinite matrices. This cone induces the partial ordering

$$A \leq B \iff \langle \eta, A\eta \rangle \leq \langle \eta, B\eta \rangle \quad \text{for every vector } \eta.$$

In the product space $B(\mathcal{H}_{AB}) = B(\mathcal{H}_A) \otimes B(\mathcal{H}_B)$, we have two natural positive cones, $B(\mathcal{H}_{AB})^+$ consists of the positive semidefinite matrices acting on $\mathcal{H}_{AB} := \mathcal{H}_A \otimes \mathcal{H}_B$, and the cone $\mathscr{S}$ consists of all operators of the form

D. Petz, *Entanglement.* In: D. Petz, Quantum Information Theory and Quantum Statistics, Theoretical and Mathematical Physics, pp. 53–71 (2008)
DOI 10.1007/978-3-540-74636-2_4

$$\sum_i A_i \otimes B_i,$$

where $A_i \in B(\mathcal{H}_A)^+$ and $B_i \in B(\mathcal{H}_B)^+$. It is obvious that $\mathcal{S} \subset B(\mathcal{H}_{AB})^+$. A state is called **separable** (or unentangled) if its density belongs to $\mathcal{S}$. The other states are the **entangled** states. Therefore the set of separable states is the convex hull of product states, or the convex hull of pure product states (since any state is the convex combination of pure states).

A pure state is separable if and only if it is a product state. Indeed, pure states are extreme points in the state space (see Lemma 2.1). If a pure state is a convex combination $\sum_i p_i P_i \otimes Q_i$ of product pure states, then this convex combination must be trivial, that is, $P \otimes Q$.

Let $(e_i)_i$ be a basis of $\mathcal{H}_A$ and $(f_j)_j$ be a basis of $\mathcal{H}_B$. Then the doubled indexed family $(e_i \otimes f_j)_{i,j}$ is a basis of $\mathcal{H}_{AB}$. (Such a basis is called **product basis**.) An arbitrary vector $\Psi \in \mathcal{H}_{AB}$ admits an expansion

$$\Psi = \sum_{i,j} c_{ij} e_i \otimes f_j \tag{4.1}$$

for some coefficients c_{ij}, $\sum_{i,j} |c_{ij}|^2 = \|\Psi\|^2$.

If $h_i = \sum_j c_{ij} f_j$, then $\Psi = \sum_i e_i \otimes h_i$; however, the vectors h_i are not orthogonal. We may want to see that a better choice of the representing vectors is possible.

Lemma 4.1. *Any unit vector $\Psi \in \mathcal{H}_{AB}$ can be written in the form*

$$\Psi = \sum_k \sqrt{p_k} g_k \otimes h_k, \tag{4.2}$$

where the vectors $g_k \in \mathcal{H}_A$ and $h_k \in \mathcal{H}_B$ are pairwise orthogonal and normalized; moreover (p_k) is a probability distribution.

Proof. We can define a conjugate-linear mapping $\Lambda : \mathcal{H}_A \to \mathcal{H}_B$ as

$$\langle \Lambda \alpha, \beta \rangle = \langle \Psi, \alpha \otimes \beta \rangle$$

for every vector $\alpha \in \mathcal{H}_A$ and $\beta \in \mathcal{H}_B$. In the computation we can use the bases $(e_i)_i$ in $\mathcal{H}_A$ and $(f_j)_j$ in $\mathcal{H}_B$. If Ψ has the expansion (4.1), then

$$\langle \Lambda e_i, f_j \rangle = \overline{c_{ij}}$$

and the adjoint Λ^* is determined by

$$\langle \Lambda^* f_j, e_i \rangle = \overline{c_{ij}}.$$

(Concerning the adjoint of a conjugate-linear mapping, see (11.4) in the Appendix.)

One can compute that the reduced density matrix ρ_A of the vector state $|\Psi\rangle$ is $\Lambda^* \Lambda$. It is enough to check that

$$\langle \Psi, |e_k\rangle\langle e_\ell|\Psi\rangle = \mathrm{Tr}\, \Lambda^* \Lambda |e_k\rangle\langle e_\ell|$$

for every k and ℓ.

Choose now the orthogonal unit vectors g_k such that they are eigenvectors of ρ_A with corresponding non-zero eigenvalues p_k, $\rho_A g_k = p_k g_k$. Then

$$h_k := \frac{1}{\sqrt{p_k}} |\Lambda g_k\rangle$$

is a family of pairwise orthogonal unit vectors. Now

$$\langle \Psi, g_k \otimes h_\ell \rangle = \langle \Lambda g_k, h_\ell \rangle = \frac{1}{\sqrt{p_\ell}} \langle \Lambda g_k, \Lambda g_\ell \rangle = \frac{1}{\sqrt{p_\ell}} \langle g_\ell, \Lambda^* \Lambda g_k \rangle = \delta_{k,\ell} \sqrt{p_\ell}$$

and we arrive at the orthogonal expansion (4.2). $\qquad\qquad\qquad\qquad\qquad$ $\square$

Expansion (4.2) is called **Schmidt decomposition**.

Let $\dim \mathscr{H}_A = \dim \mathscr{H}_B = n$. A pure state $|\Phi\rangle\langle\Phi|$ on the Hilbert space $\mathscr{H}_A \otimes \mathscr{H}_B$ is called **maximally entangled** if the following equivalent conditions hold:

- The reduced densities are maximally mixed states.
- When the vector $|\Phi\rangle$ is written in the form (4.2), then $p_k = n^{-1}$ for every $1 \le k \le n$.
- There is a product basis such that $|\Phi\rangle$ is complementary to it.

The density matrix of a maximally entangled state on $\mathbb{C}^n \otimes \mathbb{C}^n$ is of the form

$$\rho = \frac{1}{\sqrt{n}} \sum_{i,j} e_{ij} \otimes e_{ij} \tag{4.3}$$

in an appropriate basis.

A common example of maximally entangled state is the **singlet state**

$$|\Phi\rangle = \frac{1}{\sqrt{2}} (|10\rangle - |01\rangle). \tag{4.4}$$

($\mathscr{H}_A = \mathscr{H}_B = \mathbb{C}^2$, which has a basis $\{|0\rangle, |1\rangle\}$.) In the singlet state there is a particular correlation between the two spins (see Example 2.6).

It is worthwhile to note that formula (4.2) also shows how to purify an arbitrary density matrix $\rho = \sum_i p_i |g_i\rangle\langle g_i|$ acting on $\mathscr{H}_A$. It is enough to choose an orthonormal family (h_i) in another Hilbert space $\mathscr{H}_B$, and (4.2) gives a pure state whose reduction is ρ. In this case $|\Psi\rangle$ is called the **purification** of ρ.

Example 4.1. Entanglement is a phenomenon appearing in case of two quantum components (Fig. 4.1). If the system is composite but one of the two components is classical, then the possible states are in the form

$$\rho^{cq} = \sum_i p_i |e_i\rangle\langle e_i| \otimes \rho_i^q, \tag{4.5}$$

where $(p_i)_i$ is a probability distribution, ρ_i^q are densities on $\mathscr{H}_q$, and $(|e_i\rangle)_i$ is a fixed basis of $\mathscr{H}_c$. (4.5) is in the convex hull of product states, therefore it is separable.

Composite systems of type classical-quantum appear often; for example, the measurement on a quantum system is described in this formalism.   □

Let $\mathscr{E} : M_n(\mathbb{C}) \to M_k(\mathbb{C})$ be a linear mapping. According to Theorem 11.24 the condition

$$\sum_{i,j} E_{ij} \otimes \mathscr{E}(E_{ij}) \geq 0 \tag{4.6}$$

is equivalent to the complete positivity of $\mathscr{E}$. Therefore the following is true.

Theorem 4.1. *The linear mapping $\mathscr{E} : M_n(\mathbb{C}) \to M_k(\mathbb{C})$ is completely positive if and only if there exists a maximally entangled state $\rho \in M_n(\mathbb{C}) \otimes M_n(\mathbb{C})$ such that*

$$(\mathrm{id}_n \otimes \mathscr{E})(\rho) \geq 0$$

holds.

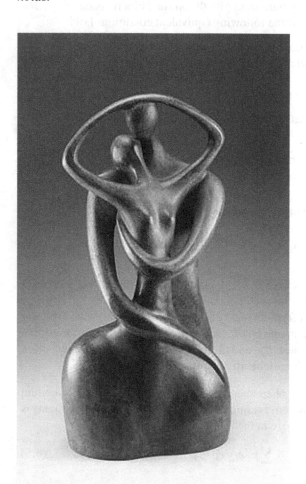

Fig. 4.1 Entanglement is a very special relation of two quantum systems. The sculpture "*Entanglement*" made by Ruth Bloch (bronze, 71 cm, 1995) might express something similar

Example 4.2. Let $\mathscr{E} : M_n(\mathbb{C}) \to M_k(\mathbb{C})$ be a channel. $\mathscr{E}$ is called **entanglement breaking** if the range of $\mathrm{id} \otimes \mathscr{E}$ contains only separable states for the identical channel $\mathrm{id} : M_n(\mathbb{C}) \to M_n(\mathbb{C})$. This condition is very restrictive; an entanglement breaking channel has the form

$$\mathscr{E}(\rho) = \sum_i \rho_i \mathrm{Tr}\, S_i \rho$$

where ρ_i is a family of states and S_i are positive matrices such that $\sum_i S_i = I$. See also Exercise 7. $\qquad\square$

Theorem 4.2. *If the state $\rho \in \mathscr{A}_{AB}$ is entangled, then there exists $W \in \mathscr{A}_{AB}^{sa}$ such that*

$$\mathrm{Tr}\, W(P \otimes Q) \geq 0$$

for all pure states $P \in \mathscr{A}_A$ and $Q \in \mathscr{A}_B$ but $\mathrm{Tr}\, W\rho < 0$.

Proof. Let $\mathscr{S}$ denote the set of separable states and assume that $\rho \notin \mathscr{S}$. Then

$$\inf\{S(\rho, D_s) : D_s \in \mathscr{S}\} > 0$$

and let D_0 be the minimizer.

It is well known that

$$\frac{\partial}{\partial \varepsilon} \log(X + \varepsilon K) = \int_0^\infty (X+t)^{-1} K (X+t)^{-1}\, dt,$$

and so

$$\frac{\partial}{\partial \varepsilon} S(\rho, (1-\varepsilon)D_0 + \varepsilon\rho') = -\mathrm{Tr}(\rho - D_0) \int_0^\infty (D_0+t)^{-1} \rho' (D_0+t)^{-1}\, dt$$

$$= 1 - \mathrm{Tr}\rho \int_0^\infty (D_0+t)^{-1} \rho' (D_0+t)^{-1}\, dt,$$

for any density ρ', both derivatives taken at $\varepsilon = 0$.

Let

$$W := I - \int_0^\infty (D_0+t)^{-1} \rho (D_0+t)^{-1}\, dt$$

and we have

$$\mathrm{Tr}\, WD_s = \frac{\partial}{\partial \varepsilon} S(\rho, (1-\varepsilon)D_0 + \varepsilon D_s) = \lim_{\varepsilon \to 0} \frac{S(\rho, (1-\varepsilon)D_0 + \varepsilon D_s) - S(\rho, D_0)}{\varepsilon} \geq 0$$

for an arbitrary $D_s \in \mathscr{S}$, since $(1-\varepsilon)D_0 + \varepsilon D_s \in \mathscr{S}$ and $S(\rho, (1-\varepsilon)D_0 + \varepsilon D_s) \geq S(\rho, D_0)$.

Due to the convexity of the relative entropy we have

$$S(\rho,(1-\varepsilon)D_0+\varepsilon\rho)-S(\rho,D_0) \le -\varepsilon S(\rho,D_0).$$

Divided by $\varepsilon > 0$ and taking the limit $\varepsilon \to 0$ we arrive at

$$\mathrm{Tr}\, W\rho \le -S(\rho,D_0) < 0.$$

$\square$

The operator W appearing in the previous theorem is called **entanglement witness**.

Example 4.3. Let

$$W := \sigma_1 \otimes \sigma_1 + \sigma_3 \otimes \sigma_3 = \begin{bmatrix} 1 & 0 & 0 & 1 \\ 0 & -1 & 1 & 0 \\ 0 & 1 & -1 & 0 \\ 1 & 0 & 0 & 1 \end{bmatrix}.$$

For any product state $\rho_1 \otimes \rho_2$, we have

$$\begin{aligned} \mathrm{Tr}\,(\rho_1 \otimes \rho_2)W &= \mathrm{Tr}\,\rho_1\sigma_1 \times \mathrm{Tr}\,\rho_2\sigma_1 + \mathrm{Tr}\,\rho_1\sigma_3 \times \mathrm{Tr}\,\rho_2\sigma_3 \\ &\le \sqrt{(\mathrm{Tr}\,\rho_1\sigma_1)^2+(\mathrm{Tr}\,\rho_1\sigma_3)^2} \times \sqrt{(\mathrm{Tr}\,\rho_2\sigma_1)^2+(\mathrm{Tr}\,\rho_2\sigma_3)^2} \\ &\le 1, \end{aligned}$$

since

$$(\mathrm{Tr}\,\rho\sigma_1)^2+(\mathrm{Tr}\,\rho\sigma_2)^2+(\mathrm{Tr}\,\rho\sigma_3)^2 \le 1$$

for any density matrix ρ. It follows that $\mathrm{Tr}\, DW \le 1$ for any separating state D on $\mathbb{C}^4$.

Consider now the density

$$\omega := \frac{1}{6}\begin{bmatrix} 2 & 0 & 0 & 2 \\ 0 & 1 & x & 0 \\ 0 & x & 1 & 0 \\ 2 & 0 & 0 & 2 \end{bmatrix}, \tag{4.7}$$

where $0 < x < 1$. Since

$$\mathrm{Tr}\,\omega W = 1 + \frac{x}{3} > 1,$$

ω must be an entangled state. $\square$

Example 4.4. An entanglement witness determines a linear functional that may separate a state from the convex hull of product state. The separation can be done by means of non-linear functionals.

Let ρ be a state and X be an observable. The **variance**

$$\delta^2(X;\rho) := \mathrm{Tr}\,\rho X^2 - (\mathrm{Tr}\,\rho X)^2 \qquad (4.8)$$

is a concave function of the variable ρ.

Let X_i be a family of observables on the system A and assume that

$$\sum_i \delta^2(X_i;\rho_A) \geq a$$

for every state ρ_A. Similarly, choose observables Y_i on the system B and let

$$\sum_i \delta^2(Y_i;\rho_B) \geq b$$

for every state ρ_B on $\mathcal{H}_B$.

Let ρ_{AB} now be a state on the composite (or bipartite) system $\mathcal{H}_A \otimes \mathcal{H}_B$. The functional

$$\psi(\rho_{AB}) := \sum_i \delta^2(X_i \otimes I + I \otimes Y_i;\rho_{AB}) \qquad (4.9)$$

is the sum of convex functionals (of the variable ρ_{AB}), therefore it is convex. For a product state $\rho_{AB} = \rho_A \otimes \rho_B$ we have

$$\sum_i \delta^2(X_i \otimes I + I \otimes Y_i;\rho_{AB}) = \sum_i \delta^2(X_i;\rho_A) + \sum_i \delta^2(Y_i;\rho_B) \geq a + b.$$

It follows that $\psi(\rho_{AB}) \geq a + b$ for every separable state ρ_{AB}. If

$$\psi(\rho_{AB}) < a + b,$$

then the state ρ_{AB} must be entangled. $\square$

Theorem 4.3. *Let ρ_{AB} be a separable state on the bipartite system $\mathcal{H}_A \otimes \mathcal{H}_B$ and let ρ_A be the reduced state. Then ρ_{AB} is more mixed than ρ_A.*

Proof. Let (r_k) be the probability vector of eigenvalues of ρ_{AB} and (q_l) is that for ρ_A. We have to show that there is a double stochastic matrix S which transform (q_l) into (r_k).

Let

$$\rho_{AB} = \sum_k r_k |e_k\rangle\langle e_k| = \sum_j p_j |x_j\rangle\langle x_j| \otimes |y_j\rangle\langle y_j|$$

be decompositions of a density matrix in terms of unit vectors $|e_k\rangle \in \mathcal{H}_A \otimes \mathcal{H}_B$, $|x_j\rangle \in \mathcal{H}_A$ and $|y_j\rangle \in \mathcal{H}_B$. The first decomposition is the Schmidt decomposition and the second one is guaranteed by the assumed separability condition. For the reduced density ρ_A we have the Schmidt decomposition and another one:

$$\rho_A = \sum_l q_l |f_l\rangle\langle f_l| = \sum_j p_j |x_j\rangle\langle x_j|,$$

where f_j is an orthonormal family in $\mathscr{H}_A$. According to Lemma 2.2 we have two unitary matrices V and W such that

$$\sum_k V_{kj}\sqrt{p_j}|x_j\rangle \otimes |y_j\rangle = \sqrt{r_k}|e_k$$

$$\sum_l W_{jl}\sqrt{q_l}|f_l\rangle = \sqrt{p_j}|x_j\rangle.$$

Combine these equations to have

$$\sum_k V_{kj} \sum_l W_{jl}\sqrt{q_l}|f_l\rangle \otimes |y_j\rangle = \sqrt{r_k}|e_k$$

and take the squared norm:

$$r_k = \sum_l \left(\sum_{j_1,j_2} \overline{V}_{kj_1} V_{kj_2} \overline{W}_{j_1 l} W_{j_2 l} \langle y_{j_1}, y_{j_2}\rangle \right) q_l$$

Introduce a matrix

$$S_{kl} = \left(\sum_{j_1,j_2} \overline{V}_{kj_1} V_{kj_2} \overline{W}_{j_1 l} W_{j_2 l} \langle y_{j_1}, y_{j_2}\rangle \right)$$

and verify that it is double stochastic.   □

Separable states behave classically in the sense that the monotonicity of the von Neumann entropy holds.

Corollary 4.1. *Let ρ_{AB} be a separable state on the bipartite system $\mathscr{H}_A \otimes \mathscr{H}_B$ and let ρ_A be the reduced state. Then $S(\rho_{AB}) \geq S(\rho_A)$.*

Proof. The statement is an immediate consequence of the theorem, since the von Neumann entropy is monotone with respect to the more mixed relation. However, I can give another proof.

First we observe that for a separable state ρ_{AB} the operator inequality

$$\rho_{AB} \leq \rho_A \otimes I_B. \tag{4.10}$$

holds. Indeed, for a product state the inequality is obvious and we can take convex combinations. Since log is matrix monotone, we have

$$-\log \rho_{AB} \geq -(\log \rho_A) \otimes I_B. \tag{4.11}$$

Taking the expectation values with respect to the state ρ_{AB}, we get $S(\rho_{AB}) \geq S(\rho_A)$.

Both proofs show that instead of the von Neumann entropy, we can take an α-entropy as well.   □

Theorem 4.4. *Let ρ_{AB} be a state on the bipartite system $\mathscr{H}_A \otimes \mathscr{H}_B$ and let ρ_A be the reduced state. If $S(\rho_{AB}) < S(\rho_A)$, then there is $\varepsilon > 0$ such that all states ω satisfying the condition $\|\omega - \rho_{AB}\| < \varepsilon$ are entangled.*

Proof. Due to the continuity of the von Neumann entropy $S(\omega) < S(\omega_A)$ holds in a neighborhood of ρ_{AB}. All these states are entangled. $\qquad\square$

Theorem 4.5. $\rho \in \mathscr{A}_{AB}$ *is separable if and only if for any* $k \in \mathbb{N}\rho$ *has a symmetric extension to* $\mathscr{A}_A^{\overset{1}{\sim}} \otimes \mathscr{A}_A^{\overset{2}{\sim}} \otimes \ldots \otimes \mathscr{A}_A^{\overset{k}{\sim}} \otimes \mathscr{A}_B$.

Proof. For a separable state the symmetric extension is easily constructed. Assume that

$$\rho = \sum_i \lambda_i A_i \otimes B_i,$$

then

$$\sum_i \lambda_i A_i \otimes A_i \otimes \cdots \otimes A_i \otimes B_i$$

is a symmetric extension.

Conversely, let ρ_n be the assumed symmetric extension and let the state φ of the infinite product algebra be limit points of ρ_n's. Since all ρ_n's are extensions of the given ρ, so is φ. According to the **quantum de Finetti theorem** (see Notes), φ is an integral of product state and so is its restriction ρ. This shows that ρ is separable. $\qquad\square$

Theorem 4.6. *Let* $\rho \in \mathscr{A}_{AB}$. *If there is a positive mapping* $\Lambda : \mathscr{A}_B \to \mathscr{A}_B$ *such that* $(\mathrm{id} \otimes \Lambda)\rho$ *is not positive, then* ρ *is entangled.*

Proof. For a product state $\rho_{\mathscr{A}} \otimes \rho_{\mathscr{B}}$ we have

$$(\mathrm{id}_{\mathscr{A}} \otimes \Lambda)(\rho_{\mathscr{A}} \otimes \rho_{\mathscr{B}}) = \rho_{\mathscr{A}} \otimes \Lambda(\rho_{\mathscr{B}}) \geq 0.$$

It follows that $(\mathrm{id}_{\mathscr{A}} \otimes \Lambda)D$ is positive when D is separable. $\qquad\square$

In place of Λ, there is no use of completely positive mapping but matrix transposition (in any basis) could be useful.

Example 4.5. Consider the state

$$\frac{1}{4} \begin{bmatrix} 1+p & 0 & 1-p & p \\ 0 & 1-p & 0 & 1-p \\ 1-p & 0 & 1-p & 0 \\ p & 1-p & 0 & 1+p \end{bmatrix},$$

where $0 \leq p \leq 1$. The partial transpose of this matrix is

$$\frac{1}{4} \begin{bmatrix} 1+p & 0 & 1-p & 0 \\ 0 & 1-p & p & 1-p \\ 1-p & p & 1-p & 0 \\ 0 & 1-p & 0 & 1+p \end{bmatrix}.$$

If this is positive, so is

$$
\begin{bmatrix} 1-p & p \\ p & 1-p \end{bmatrix}.
$$

For $1/2 < p$ this matrix is not positive, therefore Theorem 4.6 tells us that the state is entangled for these values of the parameter. □

Example 4.6. Let $|\Psi\rangle\langle\Psi| \in \mathbb{C}^2 \otimes \mathbb{C}^2$ be a maximally entangled state and τ be the tracial state. Since τ is separable

$$
\rho_p := p|\Psi\rangle\langle\Psi| + (1-p)\tau \qquad (0 \le p \le 1) \tag{4.12}
$$

is an interpolation between an entangled state and a separable state. ρ_p is called **Werner state**; its eigenvalues are

$$
p + \frac{1-p}{4}, \frac{1-p}{4}, \frac{1-p}{4}, \frac{1-p}{4} \tag{4.13}
$$

and the eigenvalues of the reduced density matrix are $(1/2, 1/2)$. (4.13) is more mixed than this pair if and only if $p \le 1/3$. Therefore, for $p > 1/3$, the state ρ_p must be entangled as per Theorem 4.3.

We can arrive at the same conclusion also from Theorem 4.6. In an appropriate basis, the matrix of ρ_p is

$$
\frac{1}{4} \begin{pmatrix} 1-p & 0 & 0 & 0 \\ 0 & 1+p & 2p & 0 \\ 0 & 2p & 1+p & 0 \\ 0 & 0 & 0 & 1-p \end{pmatrix}. \tag{4.14}
$$

The partial transpose of this matrix is

$$
\frac{1}{4} \begin{pmatrix} 1-p & 0 & 0 & 2p \\ 0 & 1+p & 0 & 0 \\ 0 & 0 & 1+p & 0 \\ 2p & 0 & 0 & 1-p \end{pmatrix} \tag{4.15}
$$

which cannot be positive when $(1-p)^2 < 4p^2$. For $p > 1/3$ this is the case and Theorem 4.6 tells us that ρ_p is entangled.

If $p = 1/3$, then ρ_p is

$$
\frac{1}{6} \begin{pmatrix} 1 & 0 & 0 & 0 \\ 0 & 2 & 1 & 0 \\ 0 & 1 & 2 & 0 \\ 0 & 0 & 0 & 1 \end{pmatrix}.
$$

and it can be shown that this is separable by presenting a decomposition. We have

$$\begin{pmatrix} 1 & 0 & 0 & 0 \\ 0 & 2 & 1 & 0 \\ 0 & 1 & 2 & 0 \\ 0 & 0 & 0 & 1 \end{pmatrix} = \frac{1}{3}\begin{pmatrix} 3 & 0 & 0 & 0 \\ 0 & 3 & 3 & 0 \\ 0 & 3 & 3 & 0 \\ 0 & 0 & 0 & 3 \end{pmatrix} + \begin{pmatrix} 0 & 0 & 0 & 0 \\ 0 & 1 & 0 & 0 \\ 0 & 0 & 1 & 0 \\ 0 & 0 & 0 & 0 \end{pmatrix}.$$

Here the first summand has a decomposition

$$\begin{pmatrix} 1 & 1 \\ 1 & 1 \end{pmatrix} \otimes \begin{pmatrix} 1 & 1 \\ 1 & 1 \end{pmatrix} + \begin{pmatrix} 1 & \varepsilon \\ \varepsilon^2 & 1 \end{pmatrix} \otimes \begin{pmatrix} 1 & \varepsilon \\ \varepsilon^2 & 1 \end{pmatrix} + \begin{pmatrix} 1 & \varepsilon^2 \\ \varepsilon & 1 \end{pmatrix} \otimes \begin{pmatrix} 1 & \varepsilon^2 \\ \varepsilon & 1 \end{pmatrix},$$

where $\varepsilon := \exp(2\pi i/3)$ and the second summand is

$$\begin{pmatrix} 1 & 0 \\ 0 & 0 \end{pmatrix} \otimes \begin{pmatrix} 0 & 0 \\ 0 & 1 \end{pmatrix} + \begin{pmatrix} 0 & 0 \\ 0 & 1 \end{pmatrix} \otimes \begin{pmatrix} 1 & 0 \\ 0 & 0 \end{pmatrix}.$$

Since separable states form a convex set, we can conclude that ρ_p is separable for every $p \le 1/3$. □

4.2 Dense Coding and Teleportation

Quantum information and classical information are very different concepts and, strictly speaking, it has no meaning to compare them. However, transmission of a single qubit can carry two bits of classical information and transmitting classical information of two bits can yield the teleportation of the state of a quantum spin. From this point of view a qubit is equivalent to two classical bits. Both protocols use a basis consisting of maximally entangled states on the four-dimensional space.

The **Bell basis** of $\mathbb{C}^2 \otimes \mathbb{C}^2$ consists of the following vectors

$$|\beta_0\rangle = \frac{1}{\sqrt{2}}(|00\rangle + |11\rangle),$$

$$|\beta_1\rangle = \frac{1}{\sqrt{2}}(|10\rangle + |01\rangle) = (\sigma_1 \otimes I)|\beta_0\rangle,$$

$$|\beta_2\rangle = \frac{i}{\sqrt{2}}(|10\rangle - |01\rangle) = (\sigma_2 \otimes I)|\beta_0\rangle,$$

$$|\beta_3\rangle = \frac{1}{\sqrt{2}}(|00\rangle - |11\rangle) = (\sigma_3 \otimes I)|\beta_0\rangle.$$

All of them give maximally entangled states of the bipartite system.

Assume that Alice wants to communicate an element of the set $\{0, 1, 2, 3\}$ to Bob and both of them have a spin. Assume that the two spins are initially in the state $|\beta_0\rangle$. Alice and Bob may follow the following protocol called **dense coding**.

1. If the number to communicate to Bob is k, Alice applies the unitary σ_k to her spin. After this the joint state of the two spins will be the kth vector of the Bell basis.

2. Alice sends her qubit to Bob and Bob will be in the possession of both spins.
3. Bob performs the measurement corresponding to the Bell basis, and the outcome will exactly be k.

Next we shall turn to the **teleportation protocol** initiated by Bennett et al. in 1993 [14]. Consider a 3-qubit system in the initial state:

$$|\psi\rangle_A \otimes |\beta_0\rangle_{XB}.$$

(So the spin A is statistically independent of the other two spins.) Assume that Alice is in a possession of the qubits A and X, and the spin B is at Bob's disposal. The aim is to convert Bob's spin into the state $|\psi\rangle$. Alice and Bob are separated; they could be far away from each other but they can communicate in a classical channel. How can this task be accomplished?

1. Alice measures the Bell basis on the spins A and X. The outcome of the measurement is an element of the set $\{0,1,2,3\}$.
2. Alice communicates this outcome to Bob in a classical communication channel.This requires the transmission of two classical bits to distinguish among $0,1,2,3$.
3. Bob applies the unitary σ_k to the state vector of spin B if the message of Alice is "k."

Then the state of spin B is the same as was the state of spin A at the beginning of the procedure.

This protocol depends on an important identity

$$|\psi\rangle_A \otimes |\beta_0\rangle_{XB} = \frac{1}{2}\sum_{k=0}^{3} |\beta_k\rangle_{AX} \otimes \sigma_k |\psi\rangle_B. \tag{4.16}$$

The measurement of Alice is described by the projections $E_i := |\beta_i\rangle\langle\beta_i| \otimes I_B$ ($0 \le i \le 3$). The outcome k appears with the probability

$$\langle \eta, (|\beta_k\rangle\langle\beta_k| \otimes I_B)\eta\rangle,$$

where η is the vector (4.16) and this is $1/4$. If the measurement gives the value k, then after the measurement the new state vector is

$$\frac{E_k\eta}{\|E_k\eta\|} = |\beta_k\rangle_{AX} \otimes \sigma_k|\psi\rangle_B.$$

When Bob applies the unitary σ_k to the state vector $\sigma_k|\psi\rangle_B$ of his spin, he really gets $|\psi\rangle_B$.

There are a few important features concerning the protocol. The actions of Alice and Bob are local, they manipulate only the spins at their disposal and they act independently of the unknown spin X. It is also important to observe that the spin A changes immediately after Alice's measurement. If this were not the case, then the procedure could be used to copy (or to clone) a quantum state which is impossible;

Wooters and Zurek argued for a "no-cloning" theorem [120]. Another price the two parties have to pay for the teleportation is the entanglement. The state of AX and B becomes separable.

The identity (4.16) implies that

$$E_k\Big(|\psi\rangle\langle\psi|_A \otimes |\beta_0\rangle\langle\beta_0|_{XB}\Big)E_k = \frac{1}{4}|\beta_0\rangle\langle\beta_0|_{AX} \otimes \big(\sigma_k|\psi\rangle\langle\psi|_B\sigma_k\big).$$

Both sides are linear in $|\psi\rangle\langle\psi|$, which can be replaced by an arbitrary density matrix ρ:

$$E_k\Big(\rho \otimes |\beta_0\rangle\langle\beta_0|_{XB}\Big)E_k = \frac{1}{4}|\beta_0\rangle\langle\beta_0|_{AX} \otimes \big(\sigma_k\rho_B\sigma_k\big), \qquad (4.17)$$

This formula shows that the teleportation protocol works for a density matrix ρ as well. The only modification is that when Bob receives the information that the outcome of Alice's measurement has been k, he must perform the transformation $D \mapsto \sigma_k D\sigma_k$ on his spin.

We can generalize the above protocol. Assume that $\mathcal{H}$ is n-dimensional and we have unitaries U_i ($1 \le i \le n^2$) such that

$$\mathrm{Tr}\, U_i^* U_j = 0 \quad \text{if} \quad i \ne j. \qquad (4.18)$$

These orthogonality relations guarantee that the operators U_i are linearly independent and they must span the linear space of all matrices.

Let $\Phi \in \mathcal{H} \otimes \mathcal{H}$ be a maximally entangled state vector and set

$$\Phi_i := (U_i \otimes I)\Phi \qquad (1 \le i \le n^2).$$

One can check that $(\Phi_i)_i$ is a basis in $\mathcal{H} \otimes \mathcal{H}$:

$$\langle(U_i \otimes I)\Phi, (U_j \otimes I)\Phi\rangle = \langle\Phi, (U_i^* U_j \otimes I)\Phi\rangle = \mathrm{Tr}\,(U_i^* U_j \otimes I)|\Phi\rangle\langle\Phi|$$

$$= \mathrm{Tr}\,\mathrm{Tr}_2\Big(U_i^* U_j \otimes I)|\Phi\rangle\langle\Phi|\Big) = \mathrm{Tr}\,(U_i^* U_j)\,\mathrm{Tr}_2|\Phi\rangle\langle\Phi|$$

$$= \frac{1}{n}\mathrm{Tr}\, U_i^* U_j.$$

Consider three quantum systems, similar to the spin-$\frac{1}{2}$ case; assume that the systems X and A are localized at Alice and B is at Bob. Each of these n-level systems are described an n-dimensional Hilbert space $\mathcal{H}$. Let the initial state be

$$\rho_A \otimes |\Phi\rangle\langle\Phi|_{XB}.$$

The density ρ is to be teleported from Alice to Bob by the following protocol:

1. Alice measures the basis $(\Phi_i)_i$ on the quantum system $A+X$. The outcome of the measurement is an element of the set $\{1,2,3,\ldots,n^2\}$.
2. Alice communicates this outcome to Bob in a classical communication channel.
3. Bob applies the state transformation $D \mapsto U_k D U_k^*$ to his quantum system B if the message of Alice is "k" ($1 \le k \le n^2$).

The measurement of Alice is described by the projections $E_i := |\Phi_i\rangle\langle\Phi_i| \otimes I_B$ ($1 \leq i \leq n^2$). The state transformation $D \mapsto U_k D U_k^*$ corresponds to the transformation $A \mapsto U_k^* A U_k$; hence to show that the protocol works we need

$$\sum_k \operatorname{Tr} E_k (\rho \otimes |\Phi\rangle\langle\Phi|) E_k (I_{AX} \otimes U_k A U_k^*) = \operatorname{Tr} \rho A \qquad (4.19)$$

for all $A \in B(\mathscr{H})$. Indeed, the left-hand side is the expectation value of A after teleportation, while the right-hand side is the expectation value in the state ρ.

Since this equation is linear both in ρ and in A, we may assume that $\rho = |\phi\rangle\langle\phi|$ and $A = |\psi\rangle\langle\psi|$. Then the right-hand side is $|\langle\psi,\phi\rangle|^2$ and the left-hand side is

$$\sum_k |\langle \phi \otimes \Phi, (U_k \otimes I)\Phi \otimes U_k^* \psi\rangle|^2 = \sum_k |\langle U_k^* \phi \otimes \Phi, \Phi \otimes U_k^* \psi\rangle|^2.$$

Since $\langle\eta_1 \otimes \Phi, \Phi \otimes \eta_2\rangle = n^{-1}\langle\eta_1, \eta_2\rangle$, the above expression equals

$$\sum_k |\langle U_k^* \phi \otimes \Phi, \Phi \otimes U_k^* \psi\rangle|^2 = \frac{1}{n^2} \sum_k |\langle U_k^* \phi, U_k^* \psi\rangle|^2 = |\langle\phi,\psi\rangle|^2.$$

This proves (4.19), which can be written more abstractly:

$$\sum_k \operatorname{Tr} (\rho \otimes \omega)(E_k \otimes T_k(A)) = \operatorname{Tr} \rho A \qquad (4.20)$$

for all $\rho, A \in B(\mathscr{H})$, where E_k is a von Neumann measurement on $\mathscr{H} \otimes \mathscr{H}$ and $T_k : B(\mathscr{H}) \to B(\mathscr{H})$ is a noiseless channel, $T_k(A) = U_k A U_k^*$ for some unitary U_k. In this style, the dense coding is the equation

$$\operatorname{Tr} \omega(T_k \otimes \operatorname{id})E_\ell = \delta_{k,\ell}. \qquad (4.21)$$

Recall that in the teleportation protocol we had a maximally entangled state $\omega = |\Phi\rangle\langle\Phi|$, a basis U_k of unitaries which determined the measurement

$$E_k = |\Phi_k\rangle\langle\Phi_k|, \qquad \Phi_i := (U_i \otimes I)\Phi$$

and the channels $T_k(A) = U_k A U_k^*$. These objects satisfy equation (4.21) as well, so the dense coding protocol works on n level systems.

Next we will see how to find unitaries satisfying the orthogonality relation (4.18).

Example 4.7. Let $e_0, e_1, \ldots, e_{n-1}$ be a basis in the n-dimensional Hilbert space $\mathscr{H}$ and let X be the unitary operator permuting the basis vectors cyclically:

$$X e_i = \begin{cases} e_{i+1} & \text{if } 0 \leq i \leq n-2, \\ e_0 & \text{if } i = n-1. \end{cases}$$

Let $q := e^{i2\pi/n}$, and another unitary be $Y e_i = q^i e_i$. It is easy to check that $YX = qXY$, or more generally the commutation relation

$$Z^k X^\ell = q^{k\ell} X^\ell Z^k \tag{4.22}$$

is satisfied. For $S_{j,k} = Z^j X^k$, we have

$$S_{j,k} = \sum_{m=0}^{n-1} q^{mj} |e_m\rangle \langle e_{m+k}| \quad \text{and} \quad S_{j,k} S_{u,v} = q^{ku} S_{j+u,k+v},$$

where the additions $m+k$, $j+u$, $k+v$ are understood modulo n. Since $\mathrm{Tr}\, S_{j,k} = 0$ when at least one of j and k is not zero, the unitaries

$$\{S_{j,k} : 0 \le j,k \le n-1\}$$

are pairwise orthogonal.

Note that $S_{j,k}$ and $S_{u,v}$ commute if $ku = jv$ mod n. These unitaries satisfy a discrete form of the **Weyl commutation relation** and the case $n = 2$ simply reduces to the Pauli matrices: $X = \sigma_1$ and $Z = \sigma_3$. (This fact motivated our notation.) $\qquad\square$

4.3 Entanglement Measures

The degree of entanglement of a state ρ_{AB} on the bipartite Hilbert space $\mathscr{H}_A \otimes \mathscr{H}_B$ is its distance from the convex set of separable states. One possibility is to use the relative entropy as a distance function. In this way, we can arrive at the concept of the **relative entropy of entanglement**:

$$E_{RE}(\rho_{AB}) := \inf\{S(\rho_{AB}\|D) : D \in \mathscr{S}^1\}$$

where D runs over the set of separable states on $\mathscr{H}_A \otimes \mathscr{H}_B$. If ρ_{AB} is faithful, then the minimizer is unique and can be called "the best separable approximation of ρ_{AB}."

Theorem 4.7. *Let ρ_{AB} be a state on a bipartite system. Then*

$$E_{RE}(\rho) \ge S(\rho_A) - S(\rho_{AB}), \quad S(\rho_B) - S(\rho_{AB}).$$

If ρ is pure, then $e_{RE}(\rho) = S(\rho_A)$.

Proof. Let D be a separable state. Then $-\log D \ge -\log D_A \otimes I_B$ (see (4.11)) and we have

$$S(\rho_{AB}\|D) = -S(\rho_{AB}) - \mathrm{Tr}\,\rho_{AB} \log D \ge -S(\rho_{AB}) - \mathrm{Tr}\,\rho_{AB}(\log D_A \otimes I_B)$$
$$= -S(\rho_{AB}) - \mathrm{Tr}\,\rho_A \log D_A = -S(\rho_{AB}) + S(\rho_A) + S(\rho_A\|D_A).$$

This gives the first inequality and the second is proven similarly.

If ρ_{AB} is pure and given by a vector $\sum_i \sqrt{p_i}|\alpha_i\rangle \otimes |\beta_i\rangle$ (where $|\alpha_i\rangle$ and $|\beta_i\rangle$ are orthogonal systems), then the state

$$D := \sum_i p_i |\alpha_i\rangle\langle\alpha_i| \otimes |\beta_i\rangle\langle\beta_i|$$

is separable and $S(\rho_{AB}||D) = S(\rho_A)$. This D is not the only minimizer, since the condition $\rho_A = D_A$ does not determine D. $\square$

The relative entropy of entanglement is a possible measure of entanglement; however, there are several different possibilities. What are the minimal requirements for the quantification of entanglement?

A bipartite entanglement measure $E(\rho)$ is a mapping from density matrices into positive real numbers: $\rho \mapsto E(\rho) \in \mathbb{R}^+$ defined for states of arbitrary bipartite systems.

1. $E(\rho) = 0$ if the state ρ is separable.
2. For pure state ρ the measure is the von Neumann entropy of the reduced density: $E(\rho) = S(\rho_A)$.

Another common example for an additional property required from an entanglement measure is the convexity which means that we require

$$E(\Sigma_i p_i\rho_i) \leq \Sigma_i p_i E(\rho_i).$$

For a state ρ the **entanglement of formation** is defined as

$$E_F(\rho) := \inf\{\Sigma_i p_i E(|\psi_i\rangle\langle\psi_i|) : \rho = \Sigma_i p_i|\psi_i\rangle\langle\psi_i|\}. \tag{4.23}$$

It follows from the definition that this is the maximal convex entanglement measure.

Let ρ_{AB} be a state on the bipartite Hilbert space $\mathscr{H}_A \otimes \mathscr{H}_B$. The **squashed entanglement** is defined as follows:

$$E_{sq}(\rho_{AB}) := \tfrac{1}{2}\inf\{S(\rho_{AC}|C) - S(\rho_{ABC}|BC) : \rho_{ABC} \text{ is an extension of}$$
$$\rho_{AB} \text{ to } \mathscr{H}_A \otimes \mathscr{H}_B \otimes \mathscr{H}_C\}$$

Due to the strong subadditivity of the von Neumann entropy, this quantity is positive. When the infimum is taken in the definition, one can restrict to finite-dimensional spaces $\mathscr{H}_C$.

Example 4.8. Assume that ρ_{AB} is a pure state. Then any extension is of the form $\rho_{AB} \otimes \rho_C$ and we have

$$S(\rho_{AC}) + S(\rho_{BC}) - S(\rho_{ABC}) - S(\rho_C)$$
$$= S(\rho_A) + S(\rho_C) + S(\rho_C) + S(\rho_B) - S(\rho_{AB}) - S(\rho_C) - S(\rho_C) = 2S(\rho_A).$$

Therefore in this case $E_{sq}(\rho_{AB}) = S(\rho_A)$. $\square$

Example 4.9. Assume that ρ_{AB} is separable. It can be shown that in this case $E_{sq}(\rho_{AB}) = 0$.

We have

$$\rho_{AB} = \sum_i p_i \rho_A^i \otimes \rho_B^i$$

for some pure states $\rho_A^i \otimes \rho_C^i$. If we choose pairwise orthogonal pure states ρ_C^i, then $\rho_{ACB} = \sum_i p_i \rho_A^i \otimes \rho_C^i \otimes \rho_B^i$ will make the infimum 0 in the definition of the squashed entanglement. $\qquad\square$

All extensions of ρ_{AB} can be obtained from a purification. Assume that ρ_{ABC_0} is a purification of ρ_{AB}, and ρ_{ABC} is an arbitrary extension. One can purify the latter state into ρ_{ABCD}, which is also a purification of ρ_{AB}. Two purifications of a state are related by an isometry or coisometry $U : \mathscr{H}_{C_0} \to \mathscr{H}_{CD}$ in a way that $\mathscr{E} := \mathrm{id}_{AB} \otimes U(\,\cdot\,)U^*$ sends ρ_{ABC_0} into ρ_{ABCD}. Therefore, we can reformulate the definition of the squashed entanglement:

$$E_{sq}(\rho_{AB}) = \tfrac{1}{2}\inf\{S(\rho_{AE}|E) - S(\rho_{ABE}|BE) : \rho_{ABE} = (\mathrm{id}_{AB} \otimes \mathscr{E})(|\Phi\rangle\langle\Phi|_{ABC}\}, \tag{4.24}$$

where the infimum is over all state transformations $\mathscr{E} : B(\mathscr{H}_{C_0}) \to B(\mathscr{H}_E)$, and $|\Phi\rangle_{ABC_0}$ is a fixed purification of ρ_{AB}.

This form of the definition allows us to prove the continuity.

Theorem 4.8. *The squashed entanglement is continuous with respect to the trace norm.*

Proof. Let ρ_{AB} and ω_{AB} be states on the bipartite Hilbert space $\mathscr{H}_A \otimes \mathscr{H}_B$. Given $\varepsilon > 0$, there is $\delta > 0$ such that $|\rho_{AB} - \omega_{AB}|_1 \leq \delta$ implies the existence of purifications ρ_{ABC_0} and ω_{ABC_0} such that $|\rho_{ABC_0} - \omega_{ABC_0}|_1 \leq \varepsilon$. Since the state transformations are contractions with respect to the trace norm, we have $|\rho_{ABE} - \omega_{ABE}|_1 \leq \varepsilon$. Now we can use Theorem 5.1 and obtain that

$$|S(\rho_{AE}|E) - S(\omega_{AE}|E)|, |S(\rho_{ABE}|BE) - S(\omega_{ABE}|BE)| \leq 4\varepsilon \log d + 2H(\varepsilon, 1-\varepsilon)$$

(with $d = \dim \mathscr{H}_A$) and this estimate implies

$$|E_{sq}(\rho_{AB}) - E_{sq}(\omega_{AB})| \leq 4\varepsilon \log d + 2H(\varepsilon, 1-\varepsilon).$$

The continuity is proven. $\qquad\square$

4.4 Notes

Theorem 4.3 was proved by Nielsen and Kempe [79].

The quantum **de Finetti theorem** is about the states of the infinite product

$$\mathscr{A} \otimes \mathscr{B} \otimes \mathscr{B} \otimes \ldots$$

which remain invariant under the fine permutations of the factors $\mathscr{B}$. Such states are called **symmetric**. The theorem obtained in [39] tells us that symmetric states are in the closed convex hull of symmetric product states. (For other generalization of the de Finetti theorem, see [83].)

Example 4.4 is from [54].

The unitaries S_{jk} in Example 4.7 give a discrete form of the **Weyl commutation relation**

$$U(h_1)U(h_2) = U(h_1 + h_2)\exp\left(i\operatorname{Im}\langle h_1, h_2\rangle\right),$$

where the unitary family is labeled by the vectors of a Hilbert space [92]. The construction in Example 4.7 is due to Schwinger [107].

The bases (e_i) and (f_j) have the property

$$|\langle e_i, f_j\rangle|^2 = \frac{1}{n}.$$

Such bases are called **complementary** or **unbiased**. They appeared in connection with the uncertainty relation (see [67], or Chap. 16 of [83]). A family of mutually unbiased bases on an n-dimensional space has cardinality at most $n+1$. It is not known if the bound is reachable for any n. (It is easy to construct $n+1$ mutually unbiased bases if $n = 2^k$.) More details about complementarity are also in [99].

Theorem 4.7 was obtained in [101].

4.5 Exercises

1. Let $e_1, e_2, \ldots, e_n$ be a basis in the Hilbert space $\mathcal{H}$. Show that the vector

$$\frac{1}{\sqrt{n}} \sum_{i,j=1}^n U_{ij} e_i \otimes e_j$$

   gives a maximally entangled state in $\mathcal{H} \otimes \mathcal{H}$ if and only if the matrix $(U_{ij})_{i,j=1}^n$ is a unitary.
2. Let Φ be a unit vector in $\mathcal{H} \otimes \mathcal{H}$ and let n be the dimension of $\mathcal{H}$. Show that Φ gives a maximally entangled state if and only if

$$\langle \eta_1 \otimes \Phi, \Phi \otimes \eta_2 \rangle = n^{-1}\langle \eta_1, \eta_2 \rangle$$

   for every vector, $\eta_1, \eta_2 \in \mathcal{H}$.
3. Let Φ be a maximally entangled state in $\mathcal{H} \otimes \mathcal{H}$ and W be a unitary on $\mathcal{H}$. Show that $(W \otimes I)\Phi$ gives a maximally entangled state.
4. Use the partial transposition and Theorem 4.6 to show that the density (4.7) is entangled.
5. Show that the matrix of an operator on $\mathbb{C}^2 \otimes \mathbb{C}^2$ is diagonal in the Bell basis if and only if it is a linear combination of the operators $\sigma_i \otimes \sigma_i$, $0 \le i \le 3$.
6. Use Theorem 4.3 to show that the density (4.7) is entangled.

7. Let $\mathcal{E} : M_n(\mathbb{C}) \to M_k(\mathbb{C})$ be a channel and assume that

$$\mathrm{id} \otimes \mathcal{E}(|\Phi\rangle\langle\Phi|)$$

is separable for a maximally entangled state $|\Phi\rangle\langle\Phi|$ in $M_n(\mathbb{C}) \otimes M_n(\mathbb{C})$, where id_n is the identity $M_n(\mathbb{C}) \to M_n(\mathbb{C})$. Show that $\mathcal{E}$ has the form

$$\mathcal{E}(\rho) = \sum_i \rho_i \,\mathrm{Tr}\, S_i \rho \qquad (4.25)$$

where ρ_i is a family of states and S_i are positive matrices such that $\sum_i S_i = I$.

8. Show that

$$\frac{1}{12} \sum_{k=1}^{3} (I_4 - \sigma_k \otimes \sigma_k)$$

is a Werner state. Use identity (2.35) to show that it is separable.

9. Show that the range of $\mathcal{E}_{p,2} \otimes \mathcal{E}_{p,2}$ does not contain an entangled state if $\mathcal{E}_{p,2} : M_2(\mathbb{C}) \to M_2(\mathbb{C})$ is the depolarizing channel and $0 \leq p \leq 1/2$.

10. Assume that three qubits are in the pure state

$$\frac{1}{\sqrt{2}}(|000\rangle + |111\rangle)$$

(called GHZ state, named after **Greeneberger–Horne–Zeilinger**). Show that all qubits are in a maximally mixed state and any two qubits are in a separating state.

11. Let $|\Psi\rangle\langle\Psi| \in B(\mathbb{C}^n \otimes \mathbb{C}^n)$ be a maximally entangled state ($n \geq 3$) and assume that for another state ρ we have $\| |\Psi\rangle\langle\Psi| - \rho \|_1 \leq 1/3$. Show that ρ is entangled.

12. Let $|\Psi\rangle\langle\Psi| \in B(\mathbb{C}^n \otimes \mathbb{C}^n)$ be a maximally entangled state and τ be the tracial state on $B(\mathbb{C}^n \otimes \mathbb{C}^n)$. When will be the **Werner state**

$$\rho_p := p|\Psi\rangle\langle\Psi| + (1-p)\tau \qquad (0 \leq p \leq 1) \qquad (4.26)$$

entangled?

13. Compute the state of the spin X after the teleportation procedure between Alice and Bob.

14. Compute the joint state of the spins A and B after the teleportation procedure.

15. Show that relation (4.18) implies the condition

$$\sum_i U_i^* A U_i = n(\mathrm{Tr}\, A) I$$

for every $A \in B(\mathcal{H})$.

Chapter 5
More About Information Quantities

In the previous chapter the Shannon entropy, its quantum analogue, the von Neumann entropy and the relative entropy were introduced and discussed. They are really the fundamental quantities and several others are reduced to them. In this chapter more complex quantities will be introduced.

5.1 Shannon's Mutual Information

The Shannon entropy measures the amount of information contained in a probability distribution. In the mathematical sense, the concept of **mutual information** is the extension of the entropy to two probability distributions. The mutual information measures the amount of information contained in a random variable about another random variable.

$$I(X \wedge Y) = \sum_{xy} p(x,y) \log \frac{p(x,y)}{p(x)q(y)}, \qquad (5.1)$$

where $p(x,y)$ is the joint distribution, p is the distribution of X and q is the distribution of Y.

The mutual information is a symmetric function of the two variables. If X and Y coincide, then $I(X \wedge Y)$ reduces to the entropy and it becomes zero when they are independent. $I(X \wedge Y)$ is a relative entropy, that of the joint distribution with respect to the product of the marginals. This observation makes sure that $I(X \wedge Y) \geq 0$.

In the case where X and Y are the source and the output of a communication channel, the quantity $I(X \wedge Y)$ measures the amount of information going through the channel. This amount cannot exceed the information of the source or that of the output. Therefore,

$$I(X \wedge Y) \leq H(X) \quad \text{and} \quad I(X \wedge Y) \leq H(Y). \qquad (5.2)$$

D. Petz, *More About Information Quantities*. In: D. Petz, Quantum Information Theory and Quantum Statistics, Theoretical and Mathematical Physics, pp. 73–82 (2008)
DOI 10.1007/978-3-540-74636-2_5 © Springer-Verlag Berlin Heidelberg 2008

The mutual information is expressed as

$$I(X \wedge Y) = H(X) - H(X|Y) \tag{5.3}$$

in terms of the conditional entropy.

The quantum analogue of the mutual information is a bit problematic. The classical formulation "*the amount of information contained in a random variable about another random variable*" is not extendable since the concept of random variable is missing in the quantum setting. However, when ρ_{AB} is a density matrix of the composite system $\mathcal{H}_A \otimes \mathcal{H}_B$, ρ_A and ρ_B are the reduced density matrices, then the relative entropy

$$S(\rho_{AB} \| \rho_A \otimes \rho_B) = S(\rho_A) + S(\rho_A) - S(\rho_{AB}) \tag{5.4}$$

resembles Shannon's mutual information and it can be interpreted as the **mutual information of the subsystems** A and B. It follows from the properties of the relative entropy that this quantity is positive and vanishes if and only if ρ_{AB} is a product.

5.2 Markov Chains

The random variables X, Y and Z (with joint distribution $p(x,y,z)$) form a **Markov chain** if

$$p(x,y,z) = p(x)p(y|x)p(z|y). \tag{5.5}$$

$X \to Y \to Z$ is a frequently used notation for a Markov chain. Markovianity implies that

$$p(z|y) = p(z|y,x). \tag{5.6}$$

If Z has the interpretation as "*future*," Y is the "*present*" and X is the "*past*," then "*future conditioned to past and present*" is the same as "*future conditioned to present*." From (5.6) we can deduce

$$H(Z|Y) = H(Z|X,Y). \tag{5.7}$$

This condition can be written in other equivalent forms as

$$H(X,Y,Z) = H(X,Y) + H(Y,Z) - H(Y), \tag{5.8}$$

or

$$H(Z|X,Y) = H(Y|X). \tag{5.9}$$

Actually both conditions are equivalent forms of the Markovianity. (5.8) tells us that in the strong subadditivity inequality of Theorem 3.2 the equality holds.

Example 5.1. Let X', Y' and Z' be random variables with joint distribution $q(x,y,z)$. One can define a probability distribution

$$p(x,y,z) = q(x,y)q(z|y) \qquad (5.10)$$

of random variables X, Y and Z. Then the joint distribution of (X,Y) is the same as that of (X',Y') and the same relation holds for (Y,Z) and (Y',Z'). Hence

$$p(z|x,y) = \frac{p(x,y,z)}{p(x,y)} = q(z|y) = p(z|y)$$

and we can conclude that $X \to Y \to Z$. This is a naturally constructed Markov chain from the arbitrary triplet (X',Y',Z').

Let $q(x,y,z)$ be the probability distribution of an arbitrary triplet (X',Y',Z') and $p(x,y,z)$ be that of $X \to Y \to Z$. Then

$$
\begin{aligned}
S(q\|p) &= \sum_{x,y,z} q(x,y,z) \log \frac{q(y)q(z|y)q(x|y,z)}{p(y)p(z|y)p(x|y)} \\
&= \sum_{x,y,z} q(x,y,z) \left(\log \frac{q(y)}{p(y)} + \log \frac{q(z|y)}{p(z|y)} + \log \frac{q(x|y,z)}{p(x|y)} \right) \\
&= S(q(y)\|p(y)) + \sum_{y,z} q(y)q(z|y) \log \frac{q(z|y)}{p(z|y)} + \sum_{x,y,z} q(x,y,z) \log \frac{q(x|y,z)}{p(x|y)} \\
&= S(q(y)\|p(y)) + \sum_{y} q(y)S(q(z|y)\|p(z|y)) + \sum_{x,y,z} q(x,y,z) \log \frac{q(x|y)q(z|x,y)}{p(x|y)q(z|y)} \\
&= S(q(y)\|p(y)) + \sum_{y} q(y)S(q(z|y)\|p(z|y)) + \sum_{y} q(y)S(q(x|y)\|p(x|y)) \\
&\quad + \sum_{x,y,z} q(x,y,z) \log \frac{q(z|x,y)}{q(z|y)}.
\end{aligned}
$$

According to our computation $S(q\|p)$ consists of three relative entropy terms depending on p and a fourth term which does not depend on p. This is

$$\sum_{x,y,z} q(x,y,z) \log \frac{q(z|x,y)}{q(z|y)} = H(X',Y',Z') + H(Y') - H(X',Y') - H(Y',Z').$$

If we want to minimize $S(q\|p)$, then we should make the relative entropy terms 0. They will vanish exactly in the case (5.10). Therefore,

$$
\begin{aligned}
\inf\{D((X',Y',Z')\|(X,Y,Z)) &: X \to Y \to Z\} \\
&= H(X',Y',Z') + H(Y') - H(X',Y') - H(Y',Z')
\end{aligned}
$$

and the minimizer is the natural Markovianization of the triplet (X',Y',Z'). $\qquad \square$

5.3 Entropy of Partied Systems

Let ρ_{AB} be a density matrix of the composite system $\mathcal{H}_A \otimes \mathcal{H}_B$ and let ρ_A be the reduced density matrix on $\mathcal{H}_A$. On the basis of analogy with classical information theory the difference

$$S(\rho_{AB}|B) := S(\rho_{AB}) - S(\rho_B)$$

is regarded as **conditional entropy** given the subsystem B. The useful identity

$$S(\rho_{AB}|B) = \log \dim \mathcal{H}_A - S(\rho_{AB} \| \tau_A \otimes \rho_B),$$

where τ_A is the tracial state, can be used to conclude a few properties of the conditional entropy. The joint convexity of the relative entropy implies that the conditional entropy is concave. The concavity yields a lower bound which is sufficient to be shown for the extreme points. Recall that for a pure state ρ_{AB}, the reduced density matrices ρ_A and ρ_B have the same non-zero eigenvalues and the same von Neumann entropies. Therefore in this case we have

$$S(\rho_{AB}|B) = -S(\rho_B) = -S(\rho_A) \geq -\log \dim \mathcal{H}_A$$

and the same lower bound holds for any state. On the other hand, the positivity of the relative entropy implies $S(\rho_{AB}|B) \leq \log \dim \mathcal{H}_A$. Therefore, we have

$$|S(\rho_{AB}|B)| \leq \log \dim \mathcal{H}_A. \tag{5.11}$$

As the example of pure state shows, the quantum conditional entropy could be negative.

Lemma 5.1. *Let ρ_{AB} and ω_{AB} be states on $\mathcal{H}_A \otimes \mathcal{H}_B$, $0 < \varepsilon < 1$ and $D_{AB} := (1 - \varepsilon)\rho_{AB} + \varepsilon\omega_{AB}$. Then*

$$|S(\rho_{AB}|B) - S(D_{AB}|B)| \leq 2\varepsilon \log d + H(\varepsilon, 1 - \varepsilon),$$

where $d = \dim \mathcal{H}_A$.

Proof. First we can benefit from the concavity of the conditional entropy

$$\begin{aligned}
S(\rho_{AB}|B) - S(D_{AB}|B) &\leq S(\rho_{AB}|B) - (1 - \varepsilon)S(\rho_{AB}|B) - \varepsilon S(\omega_{AB}|B) \\
&= \varepsilon(S(\rho_{AB}|B) - S(\omega_{AB}|B)) \\
&\leq 2\varepsilon \log d
\end{aligned}$$

and this gives immediately a good upper bound.

To have a lower bound, we should estimate some von Neumann entropies. From the concavity of the von Neumann entropy

$$S(D_B) \geq (1 - \varepsilon)S(\rho_B) + \varepsilon S(\omega_B).$$

and we have the complementary inequality

$$S(D_{AB}) \leq (1-\varepsilon)S(\rho_{AB}) + \varepsilon S(\omega_{AB}) + H(\varepsilon, 1-\varepsilon),$$

see Theorem 3.7. Putting together, we obtain

$$
\begin{aligned}
S(\rho_{AB}|B) - S(D_{AB}|B) &= S(\rho_{AB}) - S(\rho_B) - S(D_{AB}) + S(D_B) \\
&\geq S(\rho_{AB}) - S(\rho_B) - (1-\varepsilon)S(\rho_{AB}) - \varepsilon S(\omega_{AB}) - H(\varepsilon, 1-\varepsilon) \\
&\quad + (1-\varepsilon)S(\rho_B) + \varepsilon S(\omega_B) \\
&= \varepsilon(S(\rho_{AB}|B) - S(\omega_{AB}|B) - H(\varepsilon, 1-\varepsilon) \\
&\geq -2\varepsilon \log d - H(\varepsilon, 1-\varepsilon)
\end{aligned}
$$

which is the lower bound. $\qquad\square$

The lemma gives a continuity estimate for the conditional entropy on the line segment connecting two states. Continuity in full generality may be reduced to the lemma.

Theorem 5.1. *Let ρ_{AB} and ω_{AB} be states on $\mathcal{H}_A \otimes \mathcal{H}_B$, and $\varepsilon := ||\rho_{AB} - \omega_{AB}||_1 < 1$. Then the following estimate holds.*

$$|S(\rho_{AB}|B) - S(\omega_{AB}|B)| \leq 4\varepsilon \log d + 2H(\varepsilon, 1-\varepsilon),$$

where $d = \dim \mathcal{H}_A$.

Remember that $\varepsilon = \mathrm{Tr}|\rho_{AB} - \omega_{AB}|$. The basic idea is to introduce the auxiliary states

$$
\begin{aligned}
D_{AB} &:= (1-\varepsilon)\rho_{AB} + |\rho_{AB} - \omega_{AB}|, \\
\sigma_{AB} &:= \varepsilon^{-1}|\rho_{AB} - \omega_{AB}|, \\
\hat{\sigma}_{AB} &:= \varepsilon^{-1}\Big((1-\varepsilon)(\rho_{AB} - \omega_{AB}) + |\rho_{AB} - \omega_{AB}|\Big).
\end{aligned}
$$

Direct computation shows that

$$D_{AB} = (1-\varepsilon)\rho_{AB} + \varepsilon\sigma_{AB} = (1-\varepsilon)\omega_{AB} + \varepsilon\hat{\sigma}_{AB};$$

Now we can estimate

$$
\begin{aligned}
|S(\rho_{AB}|B) - S(\omega_{AB}|B)| &\leq |S(\rho_{AB}|B) - S(D_{AB}|B)| + |S(\omega_{AB}|B) - D_{AB}|B) \\
&\leq 4\varepsilon \log d + 2H(\varepsilon, 1-\varepsilon),
\end{aligned}
$$

where the first inequality is obvious and the second comes from the lemma. $\qquad\square$

Note that the particular case when the B component is missing gives an estimate for the continuity of the von Neumann entropy.

5.4 Strong Subadditivity of the von Neumann Entropy

The von Neumann entropy of a density matrix ρ is defined by $S(\rho) := -\text{Tr}(\rho \log \rho)$.
Suppose ρ_{ABC} is a density matrix for a system with three components, A, B and C,
The **strong subadditivity** inequality states that

$$S(\rho_{ABC}) + S(\rho_B) \leq S(\rho_{AB}) + S(\rho_{BC}), \tag{5.12}$$

where notations like ρ_B denote the appropriate reduced density matrices of $\rho_{ABC} \in$
$B(\mathcal{H}_A \otimes \mathcal{H}_B \otimes \mathcal{H}_C)$.

The strong subadditivity inequality appears quite mysterious at first sight. Some
intuition is gained by reexpressing strong subadditivity in terms of the **conditional
entropy** $S(\rho_{AB}|B) := S(\rho_{AB}) - S(\rho_B)$. Classically, when the von Neumann entropy
is replaced by the Shannon entropy function, the conditional entropy has an interpretation as the average uncertainty about system A, given knowledge of system B
(see (3.5)). Although this interpretation is more problematic in the quantum case —
for one thing, the quantum conditional entropy can be negative! — it can still be
useful for developing intuition and suggesting results. In particular, we can see that
strong subadditivity may be recast in the equivalent form

$$S(\rho_{ABC}|BC) \leq S(\rho_{AB}|B). \tag{5.13}$$

That is, strong subadditivity expresses the intuition that our uncertainty about system A when systems B and C are known is not more than when only system B is
known. (Inequality (5.13) is to be compared with the classical situation (3.8).)

Our proof strategy is to show that strong subadditivity is implied by a related
result, the **monotonicity of the relative entropy**. To see that monotonicity of the
relative entropy implies strong subadditivity, we can reexpress strong subadditivity
in terms of the relative entropy, using the identity

$$S(\rho_{AB}|A) = \log d_B - S(\rho_{AB}\|\rho_A \otimes d_B^{-1} I_B), \tag{5.14}$$

where d_B is the dimension of $\mathcal{H}_B$. Proving this identity is a straightforward application of the definitions. Using this identity we may recast the conditional entropic
form of strong subadditivity, (5.13), as an equivalent inequality between relative
entropies:

$$S(\rho_{AB}\|d_A^{-1} I_A \otimes \rho_B) \leq S(\rho_{ABC}\|d_A^{-1} I_A \otimes \rho_{BC}) \tag{5.15}$$

This inequality obviously follows from the monotonicity of the relative entropy, and
thus strong subadditivity also follows from the monotonicity of the relative entropy.

Another proof can be given using Lieb's extension of the Golden–Thompson
inequality, (Theorem 11.29).

The operator

$$\exp(\log \rho_{AB} - \log \rho_B + \log \rho_{BC})$$

is positive and can be written as $\lambda \omega$ for a density matrix ω. We have

$$S(\rho_{AB}) + S(\rho_{BC}) - S(\rho_{ABC}) - S(\rho_B) \tag{5.16}$$
$$= \operatorname{Tr} \rho_{ABC} \left(\log \rho_{ABC} - (\log \rho_{AB} - \log \rho_B + \log \rho_{BC}) \right)$$
$$= S(\rho_{ABC} \| \lambda \omega) = S(\rho_{ABC} \| \omega) - \log \lambda$$

Therefore, $\lambda \leq 1$ implies the positivity of the left-hand-side (and the strong subadditivity). Due to Theorem 11.29, we have

$$\operatorname{Tr} \exp(\log \rho_{AB} - \log \rho_B + \log \rho_{BC}) \leq \int_0^\infty \operatorname{Tr} \rho_{AB}(tI + \rho_B)^{-1} \rho_{BC}(tI + \rho_B)^{-1} \, dt$$

Applying the partial traces we have

$$\operatorname{Tr} \rho_{AB}(tI + \rho_B)^{-1} \rho_{BC}(tI + \rho_B)^{-1} = \operatorname{Tr} \rho_B(tI + \rho_B)^{-1} \rho_B(tI + \rho_B)^{-1}$$

and that can be integrated out. Hence

$$\int_0^\infty \operatorname{Tr} \rho_{AB}(tI + \rho_B)^{-1} \rho_{BC}(tI + \rho_B)^{-1} \, dt = \operatorname{Tr} \rho_B = 1.$$

and $\lambda \leq 1$. This gives the strong subadditivity. If the equality holds in (5.12), then $\exp(\log \rho_{AB} - \log \rho_B + \log \rho_{BC})$ is a density matrix and

$$S(\rho_{ABC} \| \exp(\log \rho_{AB} - \log \rho_B + \log \rho_{BC})) = 0$$

implies

$$\log \rho_{ABC} = \log \rho_{AB} - \log \rho_B + \log \rho_{BC}. \tag{5.17}$$

This is the necessary and sufficient condition for the equality.

A more general form of the subadditivity is discussed in Chap. 9.

5.5 The Holevo Quantity

Let $\mathcal{E} : B(\mathcal{H}) \to B(\mathcal{K})$ be a state transformation. The **Holevo quantity** is defined as

$$I((p_x), (\rho_x), \mathcal{E}) := S(\mathcal{E}(\rho)) - \sum_i p_x S(\mathcal{E}(\rho_x)), \tag{5.18}$$

where $\rho = \sum_x p_x \rho_x$, (p_x) is a probability distribution and ρ_x are densities. Equivalently we have

$$I((p_x), (\rho_x), \mathcal{E}) = \sum_x p(x) S(\mathcal{E}(\rho_x) \| \mathcal{E}(\rho)). \tag{5.19}$$

From this relative entropic form some properties are observed more easily. For example,

$$I((p_x),(\rho_x),\mathscr{E}) \geq I((p_x),(\rho_x),\mathscr{E}' \circ \mathscr{E})$$

is a sort of monotonicity.

The Holevo quantity is related to transmission of classical information through the quantum channel $\mathscr{E}$. Let $\mathscr{X}$ be the input alphabet and $\mathscr{Y}$ be the output alphabet. Assume a quantum channel $\mathscr{E} : B(\mathscr{H}) \to B(\mathscr{K})$ is at our disposal and we want to use it to transmit classical information. Choose a input distribution $p(x)$ on $\mathscr{X}$ and code an input character $x \in \mathscr{X}$ by a state $\rho_x \in B(\mathscr{H})$. The receiver performs a measurement on the output system $\mathscr{K}$. This means that for every $y \in \mathscr{Y}$ a positive operator $F_y \in B(\mathscr{K})$ is given and $\sum_y F_y = I_{\mathscr{K}}$. The coding and the measurement makes a classical channel from $\mathscr{E}$, the probability that $y \in \mathscr{Y}$ is observed when $x \in \mathscr{X}$ was sent is

$$T_{yx} = \mathrm{Tr}\,\rho_x F_y.$$

The amount of classical information going through the channel is bounded by the Holevo quantity:

$$I(X \wedge Y) \leq I((p_x),(\rho_x),\mathscr{E}). \tag{5.20}$$

Holevo proved this inequality in 1973, when the concept of quantum relative entropy was not well understood yet. The inequality is written as

$$\sum_x p(x)D(\mathrm{Tr}\,\mathscr{E}(\rho_x)F_y\|\mathrm{Tr}\,\mathscr{E}(\rho)F_y) \leq \sum_x p(x)S(\mathscr{E}(\rho_x)\|\mathscr{E}(\rho)).$$

This holds, since

$$D(\mathrm{Tr}\,\mathscr{E}(\rho_x)F_y\|\mathrm{Tr}\,\mathscr{E}(\rho)F_y) \leq S(\mathscr{E}(\rho_x)\|\mathscr{E}(\rho))$$

for every x as per Example 3.7.

Equation (5.20) bounds the performance of the detecting scheme. One can see that in most cases the bound cannot be achieved.

5.6 The Entropy Exchange

Let $\mathscr{E} : B(\mathscr{H}) \to B(\mathscr{K})$ be a state transformation with Kraus representation

$$\mathscr{E}(D) = \sum_i V_i D V_i^* \qquad (V_i : \mathscr{H} \to \mathscr{K}).$$

For any density matrix ρ, the matrix

$$\left(\mathrm{Tr}\,V_i \rho V_j^*\right)_{i,j=1}^n \tag{5.21}$$

is positive and has trace 1. Its von Neumann entropy is called **entropy exchange**:

$$S(\rho;\mathscr{E}) := S\left(\mathrm{Tr}\, V_i \rho V_j^*\right) \tag{5.22}$$

Although the Kraus representation is not unique, this quantity is well defined. Assume that

$$\mathscr{E}(D) = \sum_i W_i D W_i^* \qquad (W_i : \mathscr{H} \to \mathscr{K})$$

is a different representation. Then there is a unitary matrix (c_{ij}) such that

$$V_i = \sum_j c_{ij} W_j,$$

see Theorem 11.25. We have

$$\mathrm{Tr}\, V_i \rho V_j^* = \mathrm{Tr}\, \textstyle\sum_k U_{ik} W_k \rho \left(\sum_l U_{jl} W_l\right)^* = \left(U\left(\mathrm{Tr}\, W_k \rho W_l^*\right) U^*\right)_{ij}$$

and this shows that the density matrices

$$\left(\mathrm{Tr}\, V_i \rho V_j^*\right)_{i,j=1}^n \qquad \text{and} \qquad \left(\mathrm{Tr}\, W_i \rho W_j^*\right)_{i,j=1}^n$$

are unitarily equivalent. Consequently, they have the same von Neumann entropy.

To indicate the reason of the terminology "entropy exchange," we can consider the unitary dilation of $\mathscr{E} : B(\mathscr{H}) \to B(\mathscr{H})$. The density $\mathscr{E}(\rho)$ is the reduction of $U(\rho_e \otimes \rho)U^*$, where U is a unitary acting on $\mathscr{H}_e \otimes \mathscr{H}$, that is,

$$\mathscr{E}(\rho) = \mathrm{Tr}_e U(\rho_e \otimes \rho)U^*.$$

Before the interaction of our system and the environment the total state is $\rho_e \otimes \rho$, whose entropy is $S(\rho)$ provided ρ_e is a pure state. After the interaction the entropy of the total system remains the same but the entropy of the reduced densities changes. The state of our system is $\mathscr{E}(\rho)$, while the density of the environment becomes (5.21). Therefore, the entropy exchange is the entropy of the environment after the interaction.

5.7 Notes

The proof of the strong subadditivity of the von Neumann entropy was first proven by Lieb and Ruskai in 1973 before the quantum relative entropy was known [72]. The proof based on Lieb's extension of the Golden–Thompson inequality is due to József Pitrik. Another proof can be given by differentiating inequality (11.26) at $r = 1$ [24]. The case of equality in SSA was first studied in 1988 by Petz [91]. Theorem 5.1 was proved by Alicki and Fannes [4].

5.8 Exercises

1. Let X, Y and Z be random variables and assume that they form a Markov chain in this order. Show that Z, Y and X form a Markov chain as well.
2. Let X and Y be random variables. Show that

$$I(X \wedge Y) = \inf\{S((X,Y)\|(X',Y')) : X' \text{ and } Y' \text{ are independent}\}.$$

3. Show that the entropy exchange $S(\rho;\mathscr{E})$ is a concave function of $\mathscr{E}$.
4. Give a proof for the strong subadditivity of the von Neumann entropy by differentiating inequality (11.26) at $r = 1$.

Chapter 6
Quantum Compression

A pure state of a quantum mechanical system is given by a unit vector of a Hilbert space. Assume that a quantum mechanical source emits a pure state $|\varphi\rangle$. If after encoding and decoding we arrive at a state $|\varphi'\rangle$ instead of $|\varphi\rangle$, our error could be small when the vectors $|\varphi\rangle$ and $|\varphi'\rangle$ are close enough. Hence to discuss the problem of source coding, or data compression in the quantum setting, we need to know how close two quantum states are.

6.1 Distances Between States

How close are two quantum states? There are many possible answers to this question. Restricting ourselves to pure states, we have to consider two unit vectors, $|\varphi\rangle$ and $|\psi\rangle$. Quantum mechanics has used the concept of transition probability, $|\langle \varphi \mid \varphi \rangle|^2$, for a long time. This quantity is phase invariant, and it lies between 0 and 1. It equals 1 if and only if the two states coincide, that is, $|\varphi\rangle$ equals $|\psi\rangle$ up to a phase.

The square root of the transition probability is called **fidelity**: $F(|\varphi\rangle, |\psi\rangle) := |\langle \varphi \mid \psi \rangle|$. Shannon used a nonnegative distortion measure, and we may regard $1 - F(|\varphi\rangle, |\psi\rangle)$ as a distortion function on quantum states.

Under a quantum operation pure states could be transformed into mixed states, hence we need extension of the fidelity:

$$F(|\varphi\rangle\langle\varphi|, \rho) = \sqrt{\langle \varphi \mid \rho \mid \varphi \rangle}, \tag{6.1}$$

or in full generality

$$F(\rho_1, \rho_2) = \mathrm{Tr}\sqrt{\rho_1^{1/2}\rho_2\rho_1^{1/2}} \tag{6.2}$$

for positive matrices ρ_1 and ρ_2. This quantity was studied by **Uhlmann** in a different context [113] and he proved a variational formula:

D. Petz, *Quantum Compression*. In: D. Petz, Quantum Information Theory and Quantum Statistics, Theoretical and Mathematical Physics, pp. 83–90 (2008)
DOI 10.1007/978-3-540-74636-2_6

Theorem 6.1. *For density matrices* ρ_1 *and* ρ_2

$$F(\rho_1, \rho_2) = \inf \left\{ \sqrt{\mathrm{Tr}\,(\rho_1 G)\,\mathrm{Tr}\,(\rho_2 G^{-1})} : 0 \le G \text{ is invertible} \right\}$$

holds.

Proof. The polar decomposition of the operator $\rho_2^{1/2}\rho_1^{1/2}$ is

$$\rho_2^{1/2}\rho_1^{1/2} = V\sqrt{\rho_1^{1/2}\rho_2\rho_1^{1/2}},$$

where V is a unitary. Hence

$$F(\rho_1, \rho_2) = \mathrm{Tr}\,V^*\rho_2^{1/2}\rho_1^{1/2} = \mathrm{Tr}\,(V^*\rho_2^{1/2}G^{-1/2})(G^{1/2}\rho_1^{1/2})$$

for any invertible positive G. Application of the Schwarz inequality (for the Hilbert–Schmidt inner product) yields $\le$ in the theorem.

To complete the proof, it is enough to see that the equality may occur for invertible ρ_1 and ρ_2. It is easy to see that $G = \rho_2^{1/2}V^*\rho_1^{-1/2}$ is positive and makes equality. $\square$

From Theorem 6.1 the symmetry of $F(\rho_1, \rho_2)$ is obvious and we can easily deduce the **monotonicity of the fidelity** under state transformation:

$$F(\mathscr{E}(\rho_1), \mathscr{E}(\rho_2))^2 \ge \mathrm{Tr}\,\mathscr{E}(\rho_1)G\,\mathrm{Tr}\,\mathscr{E}(\rho_2)G^{-1} - \varepsilon$$
$$\ge \mathrm{Tr}\,\rho_1\mathscr{E}^*(G)\,\mathrm{Tr}\,\rho_2\mathscr{E}^*(G^{-1}) - \varepsilon,$$

where $\mathscr{E}^*$ is the adjoint of $\mathscr{E}$ with respect to the Hilbert–Schmidt inner product, $\varepsilon > 0$ is arbitrary and G is chosen to be appropriate. It is well known that $\mathscr{E}^*$ is unital and positive, hence $\mathscr{E}^*(G)^{-1} \ge \mathscr{E}^*(G^{-1})$ (see Theorem 11.23).

$$\mathrm{Tr}\,\rho_1\mathscr{E}^*(G)\,\mathrm{Tr}\,\rho_2\mathscr{E}^*(G^{-1}) \ge \mathrm{Tr}\,\rho_1\mathscr{E}^*(G)\,\mathrm{Tr}\,\rho_2\mathscr{E}^*(G)^{-1}$$
$$\ge F(\rho_1, \rho_2)^2.$$

In this way the monotonicity is concluded.

Theorem 6.2. *For a state transformation* $\mathscr{E}$ *the inequality*

$$F(\mathscr{E}(\rho_1), \mathscr{E}(\rho_2)) \ge F(\rho_1, \rho_2)$$

holds.

From the definition (6.2) one observes that $F(\rho_1, \rho_2)$ is concave in ρ_2. (Remember that $\sqrt{t}$ is operator concave, see Example 11.24.) However, the monotonicity gives that $F(\rho_1, \rho_2)$ is **jointly concave** as well. Consider the state transformation

$$\mathscr{E} : \begin{bmatrix} A & B \\ C & D \end{bmatrix} \mapsto A + D$$

Then

$$\lambda F(\rho_1,\rho_2)+(1-\lambda)F(\rho_1',\rho_2') = F\left(\begin{bmatrix} \lambda\rho_1 & 0 \\ 0 & (1-\lambda)\rho_1' \end{bmatrix}, \begin{bmatrix} \lambda\rho_2 & 0 \\ 0 & (1-\lambda)\rho_2' \end{bmatrix}\right)$$
$$\leq F(\lambda\rho_1+(1-\lambda)\rho_1',\lambda\rho_2+(1-\lambda)\rho_2')$$

as an application of monotonicity and the concavity is obtained.

Another remarkable operational formula is

$$F(\rho_1,\rho_2) = \max \{|\langle\psi_1|\psi_2\rangle| : \mathscr{E}(|\psi_1\rangle\langle\psi_1|) = \rho_1, \tag{6.3}$$
$$\mathscr{E}(|\psi_2\rangle\langle\psi_2|) = \rho_2 \text{ for some state transformation } \mathscr{E}\}.$$

This variational expression reduces the understanding of the fidelity of arbitrary states to the case of pure states. The monotonicity property is implied by this formula easily.

Convergence in fidelity is equivalent to convergence in trace norm: $F(\rho_n,\rho_n') \to 1$ if and only if $\mathrm{Tr}\,|\rho_n - \rho_n'| \to 0$. This property of the fidelity is a consequence of the inequalities

$$1 - F(\rho_1,\rho_2) \leq \frac{1}{2}\mathrm{Tr}\,|\rho_1 - \rho_2| \leq \sqrt{1 - F(\rho_1,\rho_2)}. \tag{6.4}$$

6.2 Reliable Compression

Let $(p_i,|\psi_i\rangle)$ be a source of pure quantum states on a Hilbert space $\mathscr{H}$. What is mean by a reliable compression of the source $(p_i,|\psi_i\rangle)$? The **compression scheme** consists of two quantum operations $\mathscr{C}^n : \mathscr{B}(\mathscr{H}^n) \to B(\mathscr{K}_n)$ and $\mathscr{D}^n : \mathscr{B}(\mathscr{K}_n) \to B(\mathscr{H}^n)$. $\mathscr{K}_n$ is a Hilbert space of dimension 2^{nR_n}. Let us assume that $\mathscr{K}_n \subset \mathscr{H}^n$ and $\mathscr{D}^n(D) = D \oplus 0$. This compression scheme is **reliable** and has **rate** R when

(i) $R_n \to R$,
(ii) $\sum_I p_I F(|\psi_I\rangle\langle\psi_I|, \mathscr{D}^n \circ \mathscr{C}^n(|\psi_I\rangle\langle\psi_I|)) \to 1$ as $n \to \infty$, where summation is over the multiindeces $I = (i_1,i_2,\ldots,i_n)$; moreover, $p_I = p_{i_1}p_{i_2}\cdots p_{i_n}$ and $|\psi_I\rangle = |\psi_{i_1}\rangle \otimes |\psi_{i_2}\rangle \otimes \ldots \otimes |\psi_{i_n}\rangle$.

The first condition tells that asymptotically 2^R dimension is used for the compression of a single emission of the source on the average. (This dimension is equivalent to the use of R qubits.) On the other hand, the second condition tells that the emitted state and the compressed one are close in the average; the expectation value of the fidelity is converging to 1. (Note that this definition of the reliable compression scheme is not the most general, since the form of $\mathscr{D}^n$ is restricted.)

The compression theorem depends on the **high probability subspace theorem** obtained by Ohya and Petz (Theorem. 1.18 in [83]).

Theorem 6.3. *Let ρ be a density matrix acting on the Hilbert space $\mathscr{H}$. Then the n-fold tensor product $\rho_n := \rho \otimes \rho \otimes \cdots \otimes \rho$ acts on the n-fold product space $\mathscr{H}_n := \mathscr{H} \otimes \mathscr{H} \otimes \cdots \otimes \mathscr{H}$. For any $1 > \varepsilon > 0$ we have*

$$\lim_{n\to\infty}\frac{1}{n}\inf\{\log\operatorname{Tr}Q_n : Q_n \text{ is a projection on } \mathscr{H}_n, \operatorname{Tr}\rho_nQ_n \geq 1-\varepsilon\} = S(\rho).$$

Roughly speaking, the theorem tells that a projection Q_n of large probability has the dimension $\exp(nS(\rho))$ at least.

Proof. First we should construct projections of high probability and of small dimension. Fix $\delta > 0$ and let $P(n,\delta)$ be the spectral projection of $-\frac{1}{n}\log\rho_n$ corresponding to the interval $(S(\rho)-\delta, S(\rho)+\delta)$. It follows that

$$(S(\rho)-\delta)P(n,\delta) \leq \left(-\frac{1}{n}\log\rho_n\right)P(n,\delta) \leq (S(\rho)+\delta)P(n,\delta)$$

and hence

$$e^{-n(S(\rho)+\delta)}P_{(n,\delta)} \leq \rho_nP(n,\delta) \leq e^{-n(S(\rho)-\delta)}P_{(n,\delta)}. \qquad (6.5)$$

From the first inequality, we can easily conclude

$$\frac{1}{n}\log\operatorname{Tr}P(n,\delta) \leq S(\rho)+\delta.$$

and $\limsup_{n\to\infty} \leq S(\rho)$ follows concerning the limit in the statement.

Let now Q_n be a projection on $\mathscr{H}_n$ such that $\operatorname{Tr} Q_n\rho_n \geq 1-\varepsilon$. This implies

$$\liminf_{n\to\infty}\operatorname{Tr}\rho_nQ_nP(n,\delta) \geq 1-\varepsilon,$$

since

$$\begin{aligned}
\operatorname{Tr}\rho_nQ_nP(n,\delta) &= \operatorname{Tr}\rho_nQ_n - \operatorname{Tr}\rho_nQ_nP(n,\delta)^\perp \\
&\geq \operatorname{Tr}\rho_nQ_n - \operatorname{Tr}\rho_nP(n,\delta)^\perp
\end{aligned}$$

and $P(n,\delta) \to I$ from the law of large numbers.

Next we can use the second inequality of (6.5) and estimate as follows:

$$\begin{aligned}
\operatorname{Tr}Q_n &\geq \operatorname{Tr}Q_nP(n,\delta) \\
&\geq \operatorname{Tr}\rho_nQ_nP(n,\delta)e^{n(S(\rho)-\delta)} \\
&= e^{n(S(\rho)-\delta)} \cdot \operatorname{Tr}\rho_nQ_nP(n,\delta)
\end{aligned}$$

and

$$\frac{1}{n}\log\operatorname{Tr}Q_n \geq S(\rho)-\delta+\frac{1}{n}\log\operatorname{Tr}\rho_nQ_nP(n,\delta).$$

When $n \to \infty$ the last term of the right-hand side converges to 0. Since $\delta > 0$ can be arbitrarily small, we can conclude that

$$\liminf_{n\to\infty}\frac{1}{n}\log\operatorname{Tr}Q_n \geq S(\rho).$$

$\square$

The positive part of **Schumacher's source coding theorem** is the following.

Theorem 6.4. *Let* $(p_i, |\psi_i\rangle)$ *be a source of pure states on a Hilbert space* $\mathscr{H}$ *and let S be the von Neumann entropy of the density matrix* $\sum_i p_i |\psi_i\rangle\langle\psi_i|$. *If* $R > S$, *then there exists a reliable compression scheme of rate R.*

Proof. Let ρ be the density matrix $\sum_i p_i, |\psi_i\rangle\langle\psi_i|$ and apply the high probability subspace theorem with a sequence $\varepsilon_n \to 0$. We can obtain a high probability subspace $\mathscr{K}_n \subset \mathscr{H}^n$ with orthogonal projection P_n such that

$$\dim \mathscr{K}_n \leq 2^{nR} \qquad \text{and} \qquad \operatorname{Tr} \rho_n P_n \to 1.$$

Fix a basis (ξ_i) in the Hilbert space $\mathscr{H}^n$ such that $\{\xi_i : i \in A\}$ is a basis of $\mathscr{K}_n$ and $\{\xi_i : i \in B\}$ is a basis of $\mathscr{K}_n^\perp$.

Next give the quantum operations $\mathscr{C}^n : \mathscr{B}(\mathscr{H}^n) \to \mathscr{B}(\mathscr{K}_n)$ and $\mathscr{D}^n : \mathscr{B}(\mathscr{K}_n) \to \mathscr{B}(\mathscr{H}^n)$. Set

$$\mathscr{C}^n(\sigma) = P_n \sigma P_n + \sum_{i \in B} X_i \sigma X_i^*,$$

where P_n is the orthogonal projection $\mathscr{H}^n \to \mathscr{K}_n$, $X_i = |\xi\rangle\langle\xi_i|$ with a fixed vector $\xi \in \mathscr{K}_n$. Since

$$\sum_{i \in B} X_i^* X_i = \sum_{i \in B} |\xi_i\rangle\langle\xi_i| = P_n^\perp,$$

$\mathscr{C}^n$ is really a state transformation.

For $\rho \in \mathscr{B}$, we have $(\mathscr{K}_n)\,\mathscr{D}^n(\rho)$ acting on $\mathscr{K}_n \subset \mathscr{H}^n$, and ρ and 0 on $\mathscr{H}^n \ominus \mathscr{K}_n$. Our task is to show that

$$F_n := \sum_I p_I F(|\psi_I\rangle\langle\psi_I|, \, \mathscr{D}^n \circ \mathscr{C}^n(|\psi_I\rangle\langle\psi_I|)$$

converges to 1. Give a lower estimate simply by neglecting the second term in the definition of $\mathscr{C}^n(\sigma)$ (see Exercise 3):

$$F_n \geq \sum_I p_I \langle\psi_I, P_n\psi_I\rangle = \operatorname{Tr}\rho_n P_n$$

that converges to 1. Hence $F_n \to 1$. □

Note that the pure state $|\psi_i\rangle$ compressed into mixed state in the scheme we have constructed. It is also remarkable that the statistical operator $\sum p_i |\psi_i\rangle\langle\psi_i|$ of the ensemble played a key role and not the ensemble itself. (Many different ensembles may have the same statistical operator.)

Now let us turn to the negative part of Schumacher's theorem.

Theorem 6.5. *Let* $(p_i, |\psi_i\rangle)$ *be a source of pure states on a Hilbert space* $\mathscr{H}$ *and let S be the von Neumann entropy of the density matrix* $\sum_i p_i |\psi_i\rangle\langle\psi_i|$. *If* $R < S$, *then reliable compression scheme of rate R does not exists.*

Proof. Assume that a reliable compression scheme of rate $R < S$ exists. Then

$$F_n := \sum_I p_I F(|\psi_I\rangle\langle\varphi_I|, \mathscr{D}^n \circ \mathscr{C}^n(|\psi_I\rangle\langle\psi_I|)$$

$$= \sum_I p_I \sqrt{\langle\psi_I|\mathscr{C}^n(|\psi_I\rangle\langle\psi_I|)|\psi_I\rangle}$$

$$\leq \sqrt{\sum_I p_I \langle\psi_I|\mathscr{C}^n(|\psi_I\rangle\langle\psi_I|)|\psi_I\rangle}$$

by concavity of the square root function. Moreover,

$$\sum_I p_I \langle\psi_I|\mathscr{C}^n(|\psi_I\rangle\langle\psi_I|)|\psi_I\rangle = \sum_I p_I \mathrm{Tr}|\psi_I\rangle\langle\psi_I|P_n \mathscr{C}^n(|\psi_I\rangle\langle\psi_I|)$$

$$\leq \sum_I p_I \mathrm{Tr}|\psi_I\rangle\langle\psi_I|P_n = \mathrm{Tr}\rho_n P_n$$

for the projection P_n of $\mathscr{H}^n$ onto $\mathscr{K}_n$. Since the compression is of rate R we have

$$\lim_n \frac{1}{n}\log\dim P_n \leq R.$$

On the other hand, the high probability subspace theorem tells us that in this case

$$\limsup_n \mathrm{Tr}\rho_n P_n \geq 1 - \varepsilon$$

is impossible for any $0 < \varepsilon < 1$. We have arrived at a contradiction with the assumption that the average fidelity is converging to 1. In fact, it has been shown that for any compression scheme of rate R the average fidelity converges to 0. $\square$

6.3 Universality

The data compression protocol depends heavily on the high probability subspace theorem. The construction of the previous section gives that the compression scheme can be constructed from the statistical operator $\sum_i p_i, |\psi_i\rangle\langle\psi_i|$. If the high probability subspace can be constructed universally for many statistical operators, then we can have a universal compression scheme as well. Since the von Neumann entropy is included in the properties of the subspace, this must be given. It turns out that this information is sufficient to construct a universal scheme for all densities of smaller entropy.

A density matrix is given by the eigenvalues λ_j and by the eigenvectors ξ_j. First we should make the large probability subspace universal in the spectrum than in the eigenbasis. When the eigenbasis is fixed, we are essentially in the framework of classical information theory and may follow the method of types developed by Csiszár and Körner (see the proof of Theorem 3.6).

Let $\xi_1, \xi_2, \ldots, \xi_k$ be the eigenvectors of the density matrix ρ and assume that $S(\rho) \leq R$. The corresponding eigenvalues $(\lambda_1, \lambda_2, \ldots, \lambda_k)$ form a probability distribution on the set $\mathcal{X} := \{1, 2, \ldots, k\}$ and $H(\lambda_1, \lambda_2, \ldots, \lambda_k) = S(\rho)$.

Let

$$A_n := \left\{ (x_1, x_2, \ldots, x_n) \in \mathcal{X}^n : H(\lambda_{x_1}, \lambda_{x_2}, \ldots \lambda_{x_n}) \leq R - k \frac{\log(n+1)}{n} \right\}.$$

We learnt in the proof of Theorem 6.6 that the cardinality of A_n is at most 2^{nR}. Therefore the dimension of the subspace $\mathcal{H}_n^0$ generated by the vectors

$$\{\xi_{x_1} \otimes \xi_{x_2} \otimes \cdots \otimes \xi_{x_n} \in \mathcal{H} : (x_1, x_2, \ldots, x_n) \in A_n\}$$

is at most 2^{nR}. Denote by P_n^0 the orthogonal projection onto the subspace $\mathcal{H}_n^0$.

When $\omega = \sum_j \kappa_j |\xi_j\rangle\langle\xi_j|$ is a density matrix such that

$$S(\omega) = H(\kappa_1, \kappa_2, \ldots, \kappa_k) < R,$$

then

$$\mathrm{Tr}\, \omega_n P_n^0 \geq 1 - (n+1)^k 2^{-nC} \qquad (C > 0) \qquad\qquad (6.6)$$

as per (3.14). So the convergence is uniform and exponential. The subspace $\mathcal{H}_n^0$ is already universal in the class of density matrices which have eigenbasis $\xi_1, \xi_2, \ldots, \xi_k$ (and von Neumann entropy smaller than R).

Let U be a unitary acting on $\mathcal{H}$. The subspace

$$\mathcal{H}_n^U := (U \otimes U \otimes \cdots \otimes U) \mathcal{H}_n^0$$

is universally good for the class of densities with eigenvectors $U\xi_1, U\xi_2, \ldots, U\xi_k$. Let $\mathcal{H}_n$ be the subspace generated by the subspaces $\mathcal{H}_n^U$ and denote by Q_n the corresponding projection. Since $\mathcal{H}_n \supset \mathcal{H}_n^U$, we still have

$$\mathrm{Tr}\, \omega_n Q_n \to 1.$$

What we should do is the estimation of the dimension of $\mathcal{H}_n$. If M denote the dimension of the linear space generated by the n-fold products $U \otimes U \otimes \ldots \otimes U$ when U runs over the unitaries on $\mathcal{H}$, then $\dim \mathcal{H}_n \leq \dim \mathcal{H}_n^0 \times M$. $U \otimes U \otimes \ldots \otimes U$ belongs to the symmetric tensor power of the space of $k \times k$ matrices. Therefore,

$$M \leq \frac{(n+k^2-1)!}{(k^2-1)! n!} \leq (n+1)^{k^2}.$$

This estimate guarantees

$$\lim_n \frac{1}{n} \log \dim Q_n \leq R.$$

The above arguments lead to the following.

Theorem 6.6. *Let $R > 0$ and $\mathcal{H}$ be a Hilbert space. There is a projection Q_n acting on the n-fold product space $\mathcal{H} \otimes \mathcal{H} \otimes \cdots \otimes \mathcal{H}$ such that*

$$\limsup_n \frac{1}{n} \log \operatorname{Tr} Q_n \leq R \tag{6.7}$$

and for any density matrix ρ acting on the Hilbert space $\mathcal{H}$ with von Neumann entropy $S(\rho) < R$

$$\lim_n \operatorname{Tr} \rho_n Q_n = 1, \tag{6.8}$$

where $\rho_n = \rho \otimes \rho \otimes \cdots \otimes \rho$ is the n-fold product state.

6.4 Notes

The operational expression (6.3) for the fidelity is from [35]. The universal compression in the setting was initiated in [65].

Let $\mathscr{A}_n$ be the n-fold tensor product of the matrix algebra M_m. When we want to work with these algebras with infinitely many values of n, or with arbitrarily large value of n, it is useful to consider the infinite tensor product $\mathscr{A}$. The finite-dimensional algebra $\mathscr{A}_n$ is identified with the subalgebra $\mathscr{A}_n \otimes \mathbb{C}I$ of $\mathscr{A}_{n+1}$. So the union $\cup_n \mathscr{A}_n$ becomes an algebra, called "local algebra." This is not complete with respect to the operator norm; its norm completion is the so-called **quasi-local algebra** $\mathscr{A}$. The quasi-local algebra plays an important role in the thermodynamics of quantum spin system, but it is the natural formalism to describe a quantum source as well. The product states are particular ergodic states of $\mathscr{A}$, and most of the results of the chapter hold for general ergodic states, (see [16, 66]).

6.5 Exercises

1. Check that (6.2) reduces to (6.1).
2. Let

$$\rho_1 := \tfrac{1}{2}(\sigma_0 + x \cdot \sigma) \quad \text{and} \quad \rho_2 := \tfrac{1}{2}(\sigma_0 + y \cdot \sigma)$$

   be the representation of the 2×2 density matrices in terms of the Pauli matrices (cf. (2.2)). Compute the trace distance $\|\rho_1 - \rho_2\|_1$ and the fidelity $F(\rho_1, \rho_2)$ in terms of the vectors x and y.
3. Use Theorem 11.9 to show that the fidelity $F(A,B) = \operatorname{Tr} \sqrt{A^{1/2} B A^{1/2}}$ of positive matrices is a monotone function of any of the two variables.

Chapter 7
Channels and Their Capacity

Roughly speaking, classical information means a 0–1 sequence and quantum information is a state of a quantum system. An information channel transfers information from the input, Alice, to the output, Bob. During the transmission, damage may occur due to the possible channel noise. One aim of information theory is to optimize the use of the transmission.

7.1 Information Channels

Let $\mathscr{X}$ and $\mathscr{Y}$ be two finite sets, $\mathscr{X}$ will be called the **input alphabet** and $\mathscr{Y}$ is the **output alphabet**. A classical channel sends a probability measure on $\mathscr{X}$ into a probability measure on $\mathscr{Y}$, or formulated in a different way: Assume that a random variable X with values in $\mathscr{X}$ is the random source and another random variable Y with values in $\mathscr{Y}$ is the output. The conditional probabilities

$$T_{yx} := \mathrm{Prob}(Y = y | X = x)$$

are characteristic for the channel. Given the **input distribution** p, the distribution of the output is

$$q(y) = \sum_x T_{yx} p(x). \tag{7.1}$$

The function $T : \mathscr{Y} \times \mathscr{X} \to [0,1]$ is a **Markov kernel**,

$$\sum_y T_{yx} = 1. \tag{7.2}$$

In the matrix notation (7.1) may be written as $q = Tp$ and condition (7.2) says that T is a column stochastic matrix.

A simple example of a channel is the **noisy typewriter**. The channel input is received exactly with probability $1/2$ and it is transformed into the next letter (cyclically) with probability $1/2$. The Markov kernel is

D. Petz, *Channels and Their Capacity*. In: D. Petz, Quantum Information Theory and Quantum Statistics, Theoretical and Mathematical Physics, pp. 91–107 (2008)
DOI 10.1007/978-3-540-74636-2_7

$$T_{yx} = \begin{cases} 1/2 & \text{if } y = x \text{ or } y = x+1, \\ 0 & \text{otherwise.} \end{cases}$$

Assume that the input has 26 symbols. It is easy to use this channel in a noiseless way; if we use only every alternate symbol, then we can transmit without error. The capacity of the noisy typewriter is at least the capacity of a noiseless channel transmitting 13 symbols.

In quantum probability it is difficult to speak about conditional probabilities; therefore the quantum channel is formulated slightly differently. The **channeling transformation** is an affine mapping of the density matrices acting on the input Hilbert space $\mathcal{H}$ into density matrices on the output Hilbert space $\mathcal{K}$. A channeling transformation extends to all linear operators; however, not all such transformations are regarded as a channel. The representation

$$\mathcal{E} : B(\mathcal{H}) \to B(\mathcal{K}), \qquad \mathcal{E}(\rho) = \sum_i V_i \rho V_i^*$$

is required with some operators $V_i : \mathcal{H} \to \mathcal{K}$ satisfying

$$\sum_i V_i^* V_i = I.$$

In other words, the quantum channel is a completely positive and trace-preserving linear mapping.

A simple quantum channel is given as

$$\mathcal{E}_p\left(\tfrac{1}{2}\sigma_0 + w \cdot \sigma\right) = p\left(\tfrac{1}{2}\sigma_0 + w \cdot \sigma\right) + (1-p)\tfrac{1}{2}\sigma_0 \qquad (0 < p < 1)$$

on 2×2 density matrices. The so-called depolarizing channel keeps the state of the qubit with probability p and moves it to the completely apolar state $\sigma_0/2$ with probability $1 - p$.

It is a crucial difference between classical and quantum mechanical channels that in the classical setting the joint distribution of the random input and output is the tool describing the transmission. In the quantum case, one cannot speak about joint distribution and even something analogous is not available.

7.2 The Shannon Capacity

When X is the input and Y is the output of a classical channel, the mutual information quantity $I(X \wedge Y)$ measures the amount of information going through the channel. The **Shannon capacity** is the maximum of all mutual informations $I(X \wedge Y)$ over distributions of the input $\mathcal{X}$:

$$C = \sup_p I(X \wedge Y). \tag{7.3}$$

(Concerning the mutual information, see (5.1).) Note that the distribution of the output and the joint distribution of the input and the output are determined by the channel when the input distribution p is given. (It can be misleading that the standard notation does not include the Markov kernel.)

Example 7.1. Assume that the **noisy typewriter** (described above) handles an alphabet of $2n$ characters. Let the input distribution be $p(x)$, then the joint distribution of the input and output is

$$p(x,y) = \frac{p(x)}{2} \quad \text{when} \quad y = x \quad \text{or} \quad y = x+1.$$

For the mutual information we obtain

$$-\sum_x \frac{p(x)+p(x-1)}{2} \log \frac{p(x)+p(x-1)}{2} - 1.$$

Since the first term is the Shannon entropy of a distribution on a $2n$-point space, the maximum of the mutual information is $\log 2n - 1 = \log n$. The maximum is attained when

$$\frac{p(x)+p(x-1)}{2} = \frac{1}{2n}$$

for every x. The capacity can be reached by using only the alternate symbols, but there are many other input distributions giving the same result. □

Example 7.2. The classical counterpart of the depolarizing channel has $\{0, 1\}$ as the input and the output alphabet. If the input is preserved by probability t, then $\text{Prob}(Y = X) = t$. This is the **symmetric binary channel** with the transfer matrix

$$\begin{bmatrix} t & 1-t \\ 1-t & t \end{bmatrix}.$$

The conditional entropy $H(Y|X)$ is $\eta(t) + \eta(1-t) = H(t, 1-t)$ and this does not depend on the input distribution $(p, 1-p)$. Therefore, the maximum of $H(Y) - H(Y|X)$ is reached when the input and the output are uniform. The capacity is

$$1 - H(t, 1-t).$$

If $t = 1/2$, then the capacity is 0. In this case the input and the output are independent and information transfer is not possible. In all other cases, the capacity is strictly positive. It will turn out that information transfer is possible with an arbitrary small probability of error.

An extension of the example is in Exercise 1. □

Assume that we have two channels given by the Markov kernels T and T'. If T is used to transfer the input X_1 and T' is used to transfer X_2 in a memoryless way, then the Markov kernel of the joint use is

$$T''_{(y_1,y_2)(x_1,x_2)} = T_{y_1x_1} T'_{y_2x_2}.$$

If the joint channel gives the output (Y_1, Y_2) from the input (X_1, X_2), then Y_1 does not depend on X_2 and Y_2 does not depend on X_1.

The next theorem tells us the additivity of the Shannon capacity.

Theorem 7.1. *In the above setting*

$$C(T'') = C(T) + C(T').$$

holds.

Proof. It is enough to show that

$$I((X_1,X_2) \wedge (Y_1,Y_2)) \leq I(X_1 \wedge Y_1) + I(X_2 \wedge Y_2).$$

This implies that $C(T'') \leq C(T) + C(T')$ and the converse is rather obvious.

$$\begin{aligned}
I((X_1,X_2) \wedge (Y_1,Y_2)) &= H(Y_1,Y_2) - H(Y_1,Y_2|X_1,X_2) \\
&= H(Y_1,Y_2) - H(Y_1|X_1,X_2) - H(Y_2|Y_1,X_1,X_2) \\
&= H(Y_1,Y_2) - H(Y_1|X_1) - H(Y_2|X_2),
\end{aligned}$$

since Y_i depends only on X_i and it is conditionally independent of everything else. From the subadditivity

$$H(Y_1,Y_2) - H(Y_1|X_1) - H(Y_2|X_2) \leq H(Y_1) - H(Y_1|X_1) + H(Y_2) - H(Y_2|X_2)$$

and this is the inequality I wanted to show.   □

Assume that the channel is used to communicate one of the messages $\{1, 2, \ldots,$ $2^m\}$ to the receiver. Since m could be very large, we can use the channel n times. The mathematical form of **memoryless** is the condition

$$T_{(y_1,y_2,\ldots,y_n)(x_1,x_2,\ldots,x_n)} = \prod_{i=1}^{n} T_{y_ix_i}.$$

A $(2^m, n)$ **code** for the channel consists of an encoding and a decoding function. The **encoding** $f_n : \{1, 2, \ldots, 2^m\} \to \mathscr{X}^n$ associates a codeword to a message. The **decoding function**

$$g_n : \mathscr{Y}^n \to \{1, 2, \ldots, 2^m\}$$

assigns a message to each channel output. Let λ_i be the probability that $1 \leq i \leq 2^m$ was sent over the channel but i was not received. Then λ_i is a **probability of error** and

$$P_e^{(n)} = \frac{1}{2^m} \sum_{i=1}^{2^m} \lambda_i$$

is the **average probability of error**.

The aim of channel coding is to keep the **transmission rate**

$$R = \frac{m}{n}$$

high and the error probability small. The transmission rate tells us how long 0–1 sequences are sent by one use of the channel.

A rate R is **achievable** if there exists a sequence of $(2^{nR}, n)$ codes such that the average probability of error tends to 0 as $n \to \infty$.

Shannon's noisy channel coding theorem tells us that the capacity is exactly the maximal achievable transmission rate.

Theorem 7.2 (The channel coding theorem). *If $R < C$ then there exists a sequence of $(2^{nR}, n)$ codes such that the average probability of error P_e^n tends to 0. Conversely, if a sequence of $(2^{nR}, n)$ codes has average probability of error tending to 0, then $R \le C$.*

The theorem remains true if the average probability of error is replaced by the maximal probability of error. The theorem has a positive and a negative part. The transmission rate $R < C$ is achievable, while $R > C$ is not.

7.3 Holevo Capacity

Let $\mathscr{H}$ and $\mathscr{K}$ be the input and output Hilbert spaces of a quantum communication system. The **channeling transformation** $\mathscr{E} : B(\mathscr{H}) \to B(\mathscr{K})$ sends density operators acting on $\mathscr{H}$ into those acting on $\mathscr{K}$. A **quantum code** is a probability distribution of finite support on the input densities. So $((p_x), (\rho_x))$ is a quantum code if (p_x) is a probability vector and ρ_x denotes states of $B(\mathscr{H})$. The quantum states ρ_x are sent over the quantum mechanical media, for example optical fiber, and yield the output quantum states $\mathscr{E}(\rho_x)$. The performance of coding and transmission is measured by the **quantum mutual information**

$$\chi((p_x), (\rho_x), \mathscr{E}) = \sum_x p_x S(\mathscr{E}(\rho_x) \| \mathscr{E}(\rho)) = S(\mathscr{E}(\rho)) - \sum_x p_x S(\mathscr{E}(\rho_x)), \quad (7.4)$$

where $\rho = \sum_x p_x \rho_x$. Taking the supremum over certain classes of quantum codes, we can obtain various capacities of the channel. Here all quantum codes are allowed.

$$C_{Ho}(\mathscr{E}) := \sup\{\chi((p_x), (\rho_x), \mathscr{E}) : ((p_x), (\rho_x)) \text{ is a quantum code }\}. \quad (7.5)$$

Since (7.4) is also called **Holevo quantity**, $C_{Ho}(\mathscr{E})$ is often referred as **Holevo capacity**. Taking the supremum over all mutual information $\chi((p_x), (\rho_x), \mathscr{E})$ such that ρ_x belongs to a fixed convex set $\mathscr{S}$ in the state space, we can get the restricted Holevo capacity $C_{Ho}^{\mathscr{S}}(\mathscr{E})$.

The Holevo capacity provides a bound on the performance when classical information is transmitted through the channel. In order to transmit classical information, the sender uses a quantum code and a measurement is performed on the output Hilbert space $\mathscr{K}$ by the receiver. The measurement of a family (A_y) of positive operators on $\mathscr{K}$ is given such that $\sum_y A_y = I$. Assume that the states ρ_x are fixed. Given the input distribution (p_x), the output distribution is

$$q(y) = \sum_x \mathrm{Tr}\, p_x \mathscr{E}(\rho_x) A_y$$

after the channeling and measurement. What we have is a classical channel with transition probabilities $T_{yx} = \mathrm{Tr}\,\mathscr{E}(\rho_x)A_y$. Shannon's mutual information of the input and the output is

$$\sum_x p_x D(\mathrm{Tr}\,\mathscr{E}(\rho_x)A_y \| \mathrm{Tr}\,\mathscr{E}(\rho)A_y), \tag{7.6}$$

where the relative entropy D is classical; we have two distributions parameterized by y.

$$D(\mathrm{Tr}\,\mathscr{E}(\rho_x)A_y \| \mathrm{Tr}\,\mathscr{E}(\rho)A_y) \leq S(\mathscr{E}(\rho_x)\|\mathscr{E}(\rho)) \tag{7.7}$$

holds due to the monotonicity of the relative entropy (applied here to the measurement channel). When we multiply this inequality by p_x and sum up over x, we can conclude as follows.

Theorem 7.3. *Shannon's mutual information (7.6) is bounded by the Holevo quantity (7.4).*

This statement was proved first by Holevo in 1973. That time relative entropy was not in use yet. The above argument shows us that the classically accessible information reaches the Holevo quantity if and only if in (7.7) equality holds for every x. (This happens very rarely, and the condition is related to sufficiency.)

In order to estimate the quantum mutual information, I introduce the concept of **relative entropy** or **divergence center**. Let $\mathscr{S}$ be a family of states and let $R > 0$. We can say that the state ρ is a divergence center for $\mathscr{S}$ with radius $\leq R$ if

$$S(\rho_s\|\rho) \leq R \qquad \text{for every } \rho_s \in \mathscr{S}. \tag{7.8}$$

In the following discussion about the geometry of relative entropy the ideas of [28] can be recognized very well.

Lemma 7.1. *Let (p_x) be a probability distribution and (ρ_x) be a family of states on the input Hilbert space of a channel $\mathscr{E}$. If D is a relative entropy center with radius $\leq R$ for $\{\mathscr{E}(\rho_x) : x\}$, then*

$$\chi((p_x),(\rho_x),\mathscr{E}) \leq R.$$

Proof. We have

$$S(\mathscr{E}(\rho_x)\|D) = -S(\mathscr{E}(\rho_x)) - \mathrm{Tr}\,\mathscr{E}(\rho_x)\log D \leq R$$

hence

$$\sum_x p_x S(\mathscr{E}(\rho_x)\|\mathscr{E}(\rho)) = -\sum_x p_x S(\mathscr{E}(\rho_x)) - \text{Tr}\,\mathscr{E}(\rho)\log\mathscr{E}(\rho)$$
$$\leq R - \text{Tr}\,\mathscr{E}(\rho)(\log\mathscr{E}(\rho) - \log D)$$
$$= R - S(\mathscr{E}(\rho)\|D).$$

It is quite clear, that the previous inequality is close to equality, if $S(\mathscr{E}(\rho_x)\|\|D)$ is close to R and $\sum_x p_x\mathscr{E}(\rho_x)$ is close to D. □

Let $\mathscr{S}$ be a family of states. The state ρ is said to be an **exact relative entropy center with radius R** if

$$R = \inf_D \sup\{S(\rho_s\|D) : \rho_s \in \mathscr{S}\} \quad \text{and} \quad \rho = \text{argmin}_D \sup\{S(\rho_s\|D) : \rho_s \in \mathscr{S}\},$$

where inf is over all densities D. (When R is finite, then there exists a minimizer, because $D \mapsto \sup\{S(\rho_s\|D) : \rho_s \in \mathscr{S})\}$ is lower semi-continuous with compact level sets, cf. Proposition 5.27 in [83].)

Lemma 7.2. *Let D_0, D_1 and ρ be states of $B(\mathscr{H})$ such that the Hilbert space $\mathscr{H}$ is finite dimensional and set $D_\lambda = (1-\lambda)D_0 + \lambda D_1$ $(0 \leq \lambda \leq 1)$. If $S(D_0\|\rho)$, $S(D_1\|\rho)$ are finite and*

$$S(D_\lambda\|\rho) \geq S(D_1\|\rho) \quad (0 \leq \lambda \leq 1)$$

then

$$S(D_1\|\rho) + S(D_0\|D_1) < S(D_0\|\rho).$$

Proof. Due to the assumption $S(D_\lambda\|\rho) < +\infty$, the kernel of D is smaller than that of D_λ. The function

$$f(\lambda) := S((D_\lambda\|\rho) - -S((1-\lambda)D_0 \mid \lambda D_1) - (1-\lambda)D_0\log\rho - \lambda D_1\log\rho$$

is convex on $[0, 1]$. Our condition is $f(\lambda) \geq f(1)$. It follows that $f'(1) \leq 0$. Hence we have

$$f'(1) = \text{Tr}\,(D_1 - D_0)(I + \log D_1) - \text{Tr}\,(D_1 - D_0)\log\rho$$
$$= S(D_1\|\rho) - S(D_0\|\rho) + S(D_0\|D_1) \leq 0.$$

This is the inequality we had to obtain.

We can note that in the differentiation of the function $f(\lambda)$ the well-known formula

$$\frac{\partial}{\partial t}\text{Tr}\,F(A + tB)\Big|_{t=0} = \text{Tr}\,(F'(A)B)$$

can be used. (See also Theorem 11.9.) □

Lemma 7.3. *Let $\{\rho_i : 1 \leq i \leq n\}$ be a finite set of states on $B(\mathcal{K})$ and assume that the Hilbert space $\mathcal{K}$ is finite dimensional. Then the exact relative entropy center is unique and it is in the convex hull of the states $\{\rho_i : i\}$.*

Proof. Let K be the (closed) convex hull of the states $\rho_1, \rho_2, \ldots, \rho_n$ and let ρ be an arbitrary state such that $S(\rho_i \| \rho) < +\infty$. There is a unique state $\rho \in \mathcal{K}$ such that $S(\rho' \| \rho)$ is minimal (where ρ' runs over K) (see Theorem 5.25 in [83]). Then $S((1-\lambda)\rho_i + \lambda\rho' \| \rho) \geq S(\rho' \| \rho)$ for every $0 \leq \lambda \leq 1$ and $1 \leq i \leq n$. It follows from the previous lemma that

$$S(\rho_i \| \rho) \geq S(\rho_i \| \rho').$$

Hence the relative entropy center of ρ_i's must be in K. The uniqueness of the exact relative entropy center follows from the fact that the relative entropy functional is strictly convex in the second variable. $\square$

The following result is due to Ohya, Petz and Watanabe [84]. It tells us that the capacity is the exact relative entropy radius of the range (on density matrices).

Theorem 7.4. *Let $\mathscr{E} : B(\mathcal{H}) \to B(\mathcal{K})$ be a channel and $\mathscr{S}$ be a convex set of states on $\mathcal{H}$. Then*

$$\sup\{\chi((p_x),(\rho_x),\mathscr{E}) : (p_x) \text{ is a probability distribution and } \rho_x \in \mathscr{S}\}$$

equals the relative entropy radius of the set $\mathscr{E}(\mathscr{S})$.

Proof. Let $((p_x), (\rho_x))$ be a quantum code such that $\rho_x \in \mathscr{S}$. Then $\chi((p_x), (\rho_x), \mathscr{E})$ is at most the relative entropy radius of $\{\mathscr{E}(\rho_x)\}$ (according to Lemma 7.2), which is obviously majorized by the relative entropy radius of $\mathscr{E}(\mathscr{S})$. Therefore, the restricted capacity does not exceed the relative entropy radius of $\mathscr{E}(\mathscr{S})$.

To prove the converse inequality, let us assume that the exact divergence radius of $\mathscr{E}(\mathscr{S})$ is larger than $t \in \mathbb{R}$. Then we can find densities $\rho_1, \rho_2, \cdots, \rho_n \in B(\mathcal{H})$ such that the exact relative entropy radius R of $\mathscr{E}(\rho_1), \ldots, \mathscr{E}(\rho_n)$ is larger than t. Lemma 7.3 tells us that the relative entropy center ρ of $\mathscr{E}(\rho_1), \ldots, \mathscr{E}(\rho_n)$ lies in their convex hull K'. By possible reordering of the states ρ_i we can achieve that

$$S(\mathscr{E}(\rho_i)\|\rho) \quad \begin{cases} = R & \text{if } 1 \leq i \leq k, \\ < R & \text{if } k < i \leq n. \end{cases}$$

Choose $\rho' \in K'$ such that $S(\rho' \| \rho)$ is minimal (ρ' is running over K'). Then

$$S(\mathscr{E}(\rho_i)\|\varepsilon\rho' + (1-\varepsilon)\rho) < R$$

for every $1 \leq i \leq k$ and $0 < \varepsilon < 1$, as per Lemma 7.2. However,

$$S(\mathscr{E}(\rho_i)\|\varepsilon\rho' + (1-\varepsilon)\rho) < R$$

for $k \leq i \leq n$ and for a small ε by a continuity argument. In this way, it can be concluded that there exists a probability distribution $(q_1, q_2, \ldots, q_k)$ such that

$$\sum_{i=1}^{k} q_i \mathscr{E}(\rho_i) = \rho, \qquad S(\mathscr{E}(\rho_i)\|\rho) = R.$$

Consider now the quantum code $((q_i), (\rho_i))$ as above and have

$$\Sigma_{i=1}^{k} q_i S(\mathscr{E}(\rho_i)\|\mathscr{E}(\Sigma_{j=1}^{k} q_j \rho_j)) = \Sigma_{i=1}^{k} q_i S(\mathscr{E}(\rho_i)\|\rho) = R.$$

So we have found a quantum code which has quantum mutual information larger than t. The channel capacity must exceed the entropy radius of the range. □

Example 7.3. Consider the symmetric binary channel on 2×2 matrices:

$$\mathscr{E}(w_0 \sigma_0 + w \cdot \sigma) = w_0 \sigma_0 + (\alpha w_1, \beta w_2, \gamma w_3) \cdot \sigma, \tag{7.9}$$

where

$$0 \le \alpha \le \beta \le \gamma$$

for the real parameters α, β, γ and

$$1 - \gamma \ge \beta - \alpha.$$

The range of the symmetric channel (acting on density matrices) has the shape of an ellipsoid.

The relative entropy radius of the range must exceed the relative entropy radius of the two-point-set:

$$\tfrac{1}{2}(\sigma_0 \pm \gamma \sigma_3) = \tfrac{1}{2}\begin{bmatrix} 1 \pm \gamma & 0 \\ 0 & 1 \mp \gamma \end{bmatrix}.$$

The relative entropy center is on the line segment connecting the two points, and due to symmetry it is the middle point, that is, the tracial state τ. Hence the relative entropy radius is

$$S\left(\tfrac{1}{2}\begin{pmatrix} 1 \pm \gamma & 0 \\ 0 & 1 \mp \gamma \end{pmatrix} \Big\| \tfrac{1}{2}\begin{pmatrix} 1 & 0 \\ 0 & 1 \end{pmatrix}\right) = \eta\left(\frac{1+\gamma}{2}\right) + \eta\left(\frac{1-\gamma}{2}\right).$$

It is easy to see that $S(\mathscr{E}(\rho)\|\tau)$ is less or equal to this relative entropy, since

$$S(\tfrac{1}{2}(\sigma_0 + w \cdot \sigma)\|\tau)$$

depends on $\|w\|$ and it is an increasing function of it. Therefore, τ is the exact relative entropy center and the capacity of the channel is

$$C_{Ho}(\mathscr{E}) = \eta\left(\frac{1+\gamma}{2}\right) + \eta\left(\frac{1-\gamma}{2}\right) \tag{7.10}$$

which is $H((1-\gamma)/2, (1+\gamma)/2)$. □

The **covariance** of a channel under some unitaries can be used to determine the relative entropy center of the range. Let $\mathscr{E} : B(\mathscr{H}) \to B(\mathscr{K})$ be a channeling transformation and $\mathscr{U}$ be a set of unitaries acting on $\mathscr{H}$. Assume that for every $U \in \mathscr{U}$ there exists a unitary $\alpha(U)$ on $\mathscr{K}$ such that

$$\mathscr{E}(U \cdot U^*) = \alpha(U)\mathscr{E}(\cdot)\alpha(U)^* \tag{7.11}$$

holds for every $U \in \mathscr{U}$. Assume that the relative entropy center of the range is $\mathscr{E}(\rho)$. Since

$$S(\mathscr{E}(\omega)\|\alpha(U)^*\mathscr{E}(\rho)\alpha(U) = S(\alpha(U)\mathscr{E}(\omega)\alpha(U)^*\|\mathscr{E}(\rho))$$
$$= S(\mathscr{E}(U\omega U^*)\|\mathscr{E}(\rho)) \leq radius$$

holds for every ω, the uniqueness of the relative entropy center implies that

$$\alpha(U)\mathscr{E}(\rho)\alpha(U)^* = \mathscr{E}(\rho) \tag{7.12}$$

for every $U \in \mathscr{U}$.

Example 7.4. The **depolarizing channel** $\mathscr{E}_{p,n} : M_n \to M_n$ is defined as

$$\mathscr{E}_{p,n}(A) = pA + (1-p)\frac{I}{n}\mathrm{Tr}A, \text{ where } -\frac{1}{n^2-1} \leq p \leq 1$$

(see Example 2.13). The covariance (7.11) holds for every unitary $U = \alpha(U)$, and (7.12) tells us the relative entropy center must be the tracial state. Moreover, $S(\mathscr{E}_{p,n}(\rho))$ has the same value for every pure state ρ and the relative entropy radius of the range is

$$S(\mathscr{E}_{p,n}(\rho)\|\tau) = \log n - H\left(p + \frac{1-p}{n}, \frac{1-p}{n}, \dots, \frac{1-p}{n}\right). \tag{7.13}$$

which is also the Holevo capacity of the channel as per Theorem 7.4.

The classical capacity of a channel is bounded by the Holevo capacity (see Theorem 7.3). In this example the classical capacity reaches the Holevo capacity.

Let $e_1, e_2, \dots, e_n$ be a basis in the Hilbert space. Use the pure state $|e_x\rangle\langle e_x|$ to encode $1 \leq x \leq n$ and consider the measurement given by the partition of unity $\{|e_x\rangle\langle e_x| : 1 \leq x \leq n\}$. Then the induced classical channel is described by the matrix

$$T_{yx} = p\delta_{x,y} + \frac{1-p}{n}.$$

The Shannon capacity is exactly (7.13) as it is stated in Exercise 7.7.   □

If the relative entropy center of the range of a state transformation $\mathscr{E} : B(\mathscr{H}) \to B(\mathscr{K})$ is the tracial trace τ, then the Holevo capacity is

$$C_{Ho}(\mathscr{E}) = \sup\{S(\mathscr{E}(\rho)\|\tau) : \rho\} = \log n - \inf\{S(\mathscr{E}(\rho)) : \rho\},$$

where n is the dimension of the Hilbert space $\mathcal{H}$ and ρ runs over the density matrices in $B(\mathcal{H})$. The quantity

$$h(\mathcal{E}) := \inf\{S(\mathcal{E}(\rho)) : \rho\} \tag{7.14}$$

is called the **minimum output entropy**.

Theorem 7.5. *Let $\mathcal{E}$, $\mathcal{F} : B(\mathcal{H}) \to B(\mathcal{K})$ be channels with finite-dimensional $\mathcal{K}$ and let $\mathcal{S}$ denote the set of all separable states on $\mathcal{H} \otimes \mathcal{H}$. Then*

$$C_{Ho}^{\mathcal{S}}(\mathcal{E} \otimes \mathcal{F}) = C_{Ho}(\mathcal{E}) + C_{Ho}(\mathcal{F}).$$

Proof. The inequality $\geq$ is obvious.

To prove the converse, take states ρ_i such that

$$S(\mathcal{E}(D)\|\mathcal{E}(\rho_1)) \leq C_{Ho}(\mathcal{E}) + \varepsilon \quad \text{and} \quad S(\mathcal{F}(D)\|\mathcal{F}(\rho_2)) \leq C_{Ho}(\mathcal{F}) + \varepsilon$$

for a given $\varepsilon > 0$ and for all densities D. It suffices to show that

$$S(\mathcal{E} \otimes \mathcal{F}(\bar{\rho})\|\mathcal{E} \otimes \mathcal{E}(\rho_1 \otimes \rho_2)) \leq C_{Ho}(\mathcal{E}) + C_{Ho}(\mathcal{F}) + 2\varepsilon$$

for every separable state $\bar{\rho}$. Assume that $\bar{\rho} = \sum_i \lambda_i \rho_i' \otimes \rho_i''$. Then

$$
\begin{aligned}
S\big((\mathcal{E} \otimes \mathcal{F})&(\textstyle\sum_i \lambda_i \rho_i' \otimes \rho_i'')\|(\mathcal{E} \otimes \mathcal{F})(\rho_1 \otimes \rho_2)\big) \\
&\leq \sum_i \lambda_i S\big((\mathcal{E} \otimes \mathcal{F})(\rho_i' \otimes \rho_i'')\|(\mathcal{E} \otimes \mathcal{F})(\rho_1 \otimes \rho_2)\big) \\
&= \sum_i \lambda_i \Big(S(\mathcal{E}(\rho_i')\|\mathcal{E}(\rho_1)) + S(\mathcal{F}(\rho_i'')\|\mathcal{F}(\rho_2))\Big) \\
&\leq \sum_i \lambda_i \Big(C_{Ho}(\mathcal{E}) + C_{Ho}(\mathcal{F}) + 2\varepsilon\Big) = C_{Ho}(\mathcal{E}) + C_{Ho}(\mathcal{F}) + 2\varepsilon,
\end{aligned}
$$

where the joint convexity was used first and then the additivity. This shows that the relative entropy radius of $(\mathcal{E} \otimes \mathcal{F})(\mathcal{S})$ cannot exceed $C_{Ho}(\mathcal{E}) + C_{Ho}(\mathcal{F})$ and the statement follows. $\square$

The theorem shows that using separable states for a quantum code the capacity of the product channel $\mathcal{E} \otimes \mathcal{F}$ will not exceed $C_{Ho}(\mathcal{E}) + C_{Ho}(\mathcal{F})$. What happens when entangled states are used? This is the **additivity question** for the capacity; it is an open problem. There is no example of channels such that the capacity is not additive.

Theorem 7.6. *Let $\mathcal{E} : B(\mathcal{H}) \to B(\mathcal{H})$ be a channel with finite-dimensional Hilbert space $\mathcal{H}$. Then the following conditions are equivalent.*

(a) $\mathcal{E}$ has the form

$$\mathcal{E}(B) = \sum_i X_i \mathrm{Tr}\,(Y_i B)$$

with some positive operators X_i and Y_i.
(b) The range of id $\otimes \mathcal{E}$ does not contain an entangled state.
(c) id $\otimes \mathcal{E}(|\Phi\rangle\langle\Phi|)$ is separable for a maximally entangled state $|\Phi\rangle\langle\Phi|$.
(d) $\mathcal{E}$ has a Kraus representation

$$\mathcal{E}(B) = \sum_j V_j^* B V_j$$

with operators V_j of rank one.

Proof. While proving that (a) implies (b), it may be assumed that $\mathcal{E}(B) = X\operatorname{Tr}(YB)$. Then

$$(\mathrm{id} \otimes \mathcal{E})(A \otimes B) = \operatorname{Tr}(YB)(A \otimes X) = \mathcal{F}(A \otimes B) \otimes X,$$

where $\mathcal{F}$ is a positive linear mapping determined by $A \otimes B \mapsto (\operatorname{Tr} BY)A$. It can be concluded that $(\mathrm{id} \otimes \mathcal{E})(W) = \mathcal{F}(W) \otimes X$ and any state in the range must be separating.

(b) $\Rightarrow$ (c) is obvious. Assume (c), this means that

$$\sum_{ij} E_{ij} \otimes \mathcal{E}(E_{ij}) = \sum_k \lambda_k p_k \otimes q_k \qquad (7.15)$$

for some projections p_k and q_k of rank one and positive numbers λ_k. (Recall that E_{ij} are the matrix units.)

From (7.15) we can deduce that

$$\mathcal{E}(E_{ij}) = \sum_k \lambda_k \langle \eta_k, E_{ij}\eta_k \rangle q_k$$

where $p_k = |\eta_k\rangle\langle\eta_k|$ for some vectors η_k. In place of E_{ij} we can put an arbitrary operator B, and can arrive at the Kraus representation stated in (d).

Finally, (d) $\Rightarrow$ (a), since

$$(|x\rangle\langle y|)^* B |x\rangle\langle y| = |y\rangle\langle y| \operatorname{Tr} B |x\rangle\langle x|,$$

and this completes the proof. $\square$

A channel satisfying the conditions of the previous theorem is called **entanglement breaking** channel. It follows from condition (a) that the range of $\mathcal{E} \otimes \mathcal{F}$ does not contain entangled states if $\mathcal{E}$ is an entanglement breaking channel and $\mathcal{F}$ is an arbitrary channel. Indeed, assume that $\mathcal{E}$ is in the form of (a) and set the positive mapping α_i as

$$\alpha_i(A \otimes B) = B \operatorname{Tr} Y_i A$$

for every i. Then

$$\mathcal{E} \otimes \mathcal{F}(W) = \sum_i X_i \otimes \mathcal{F}(\alpha_i(W))$$

and all states in the range are separable. Now application of Theorem 7.5 gives the following.

Theorem 7.7. *For an entanglement breaking channel $\mathscr{E} : B(\mathscr{H}) \rightarrow B(\mathscr{H})$ and for an arbitrary channel $\mathscr{F} : B(\mathscr{H}) \rightarrow B(\mathscr{H})$, the additivity of the capacity holds, that is,*

$$C_{Ho}(\mathscr{E} \otimes \mathscr{F}) = C_{Ho}(\mathscr{E}) + C_{Ho}(\mathscr{F}).$$

The lack of entanglement makes the situation close to the classical case.

Example 7.5. In this example, I want to compute the Holevo capacity of the square of the depolarizing channel (see (2.32)). The majorization of density matrices will be used.

By easy computation we have

$$(\mathscr{E}_p \otimes \mathscr{E}_p)(X) = p^2 X + \frac{p(1-p)}{n} I \otimes \mathrm{Tr}_1 X + \frac{p(1-p)}{n} \mathrm{Tr}_2 X \otimes I + \frac{(1-p)^2 \mathrm{Tr} X}{n^2} I \otimes I.$$

Let us study the action of $(\mathscr{E}_p \otimes \mathscr{E}_p)$ on projections of rank 1. If X is such a projection, then $D_1 := \mathrm{Tr}_2 X$ and $D_2 := \mathrm{Tr}_1 X$ have the same spectrum. In particular,

$$(\mathscr{E}_p \otimes \mathscr{E}_p)(R \otimes R) = p^2 R \otimes R + \frac{p(1-p)}{n} I \otimes R + \frac{p(1-p)}{n} R \otimes I + \frac{(1-p)^2}{n^2} I \otimes I.$$

To show that

$$S\big((\mathscr{E}_p \otimes \mathscr{E}_p)(Q)\big) \geq S\big((\mathscr{E}_p \otimes \mathscr{E}_p)(R \otimes R)\big),$$

the **majorization** relation $\prec$ is used. Recall that $A \prec B$ means that $\sigma_k(A) \leq \sigma_k(B)$ for every k (see the Appendix). It can be shown that

$$np\, Q + (1-p)I \otimes D_2 + (1-p)D_1 \otimes I \prec npR \otimes R + (1-p)I \otimes R + (1-p)R \otimes I. \tag{7.16}$$

The right-hand side of (7.16) consists of commuting operators and it is easy to compute the eigenvalues: $np + 2(1-p)$ with multiplicity 1, $(1-p)$ with multiplicity $2(n-1)$, and 0 with multiplicity $(n-1)^2$. Therefore, $\sigma_k = np + (k+1)(1-p)$ for $1 \leq k \leq 2n-1$ and $\sigma_k = np + 2n(1-p)$ for $2n \leq k \leq n^2$.

To analyze the left-hand side of (7.16), the decreasingly ordered eigenvalues of D_1 and D_2 are denoted by $\lambda_1, \lambda_2, \ldots, \lambda_n$. Then the eigenvalues of $I \otimes D_2 + D_1 \otimes I$ are $\lambda_i + \lambda_j (1 \leq i, j \leq n)$. The largest eigenvalue of the left-hand side of (7.16) is at most $np + 2\lambda_1(1-p)$, which cannot exceed the largest eigenvalue of the right-hand side of (7.16); in fact, it is strictly smaller when Q is not a product. To estimate the sums of the largest eigenvalues, we can use the fact

$$\sigma_k(npQ + (1-p)I \otimes D_2 + (1-p)D_1 \otimes I) \leq np + (1-p)\sigma_k(I \otimes D_2 + D_1 \otimes I)$$

and separate two cases. If $\lambda_i \leq 1/2$ for all i, then

$$\sigma_k(I \otimes D_2 + D_1 \otimes I) \le k \qquad (1 \le k \le 2n-1)$$

and if $\lambda_1 > 1/2$, then

$$\sigma_k(I \otimes D_2 + D_1 \otimes I) \le 2\lambda_1 + (k-1) \le k+1 \qquad (1 \le k \le 2n-1).$$

We still need to treat $2n \le k \le n^2$. Then

$$\sigma_k(npQ + (1-p)I \otimes D_2 + (1-p)D_1 \otimes I)$$
$$\le \mathrm{Tr}\,(npQ + (1-p)I \otimes D_2 + (1-p)D_1 \otimes I) = np + 2n(1-p) = \sigma_k.$$

This completes the majorization.

Our calculation gives that

$$S\big((\mathscr{E}_p \otimes \mathscr{E}_p)(Q), \tau \otimes \tau\big) \le S\big((\mathscr{E}_p \otimes \mathscr{E}_p)(R \otimes R), \tau \otimes \tau\big) = 2S\big(\mathscr{E}_p(R), \tau\big).$$

Since the latter is twice the capacity of the depolarizing channel $\mathscr{E}_p$, in this case the capacity is additive. □

Theorem 7.8. *Assume that* $\mathscr{E} : B(\mathscr{H}) \to B(\mathscr{K})$ *is a channeling transformation and* $C := ((p_x), (\rho_x))$ *is a quantum code with Holevo quantity*

$$\chi := \chi((p_x), (\rho_x), \mathscr{E}) = \sum_x p_x S(\mathscr{E}(\rho_x) \| \mathscr{E}(\rho)),$$

where $\rho = \sum_x p_x \rho_x$. *For every* $\varepsilon > 0$, *there is a natural number* n *and a measurement* (A_y) *on* $\mathscr{K}^{n\otimes}$ *such that the Shannon's mutual information* M *of the classical channel induced by the product quantum code* $C^{n\otimes}$ *and the measurement satisfies*

$$M/n \ge \chi - \varepsilon.$$

This means that the Holevo bound can be reached by the classical mutual information pro channel use asymptotically.

7.4 Classical-quantum Channels

Let $\mathscr{X}$ be a finite set, called "classical alphabet" or "classical input." Each $x \in \mathscr{X}$ is transformed into an output quantum state $\rho_x \in B(\mathscr{K})$, and the mapping $W : x \mapsto \rho_x$ is a **classical-quantum channel**.

The classical-quantum channel shows up if classical information is transferred through a quantum channel $\mathscr{E} : B(\mathscr{H}) \to B(\mathscr{K})$. When $x \in \mathscr{X}$ is signaled by a state $\sigma_x \in B(\mathscr{H})$, which is transformed into $\mathscr{E}(\sigma_x) \in B(\mathscr{K})$, then $W : x \mapsto \mathscr{E}(\sigma_x)$ is a classical-quantum channel.

If a classical random variable X is chosen to be the input, with probability a distribution p on $\mathscr{X}$, then the corresponding output is the quantum state $\rho_X := \sum_{x \in \mathscr{X}} p(x)\rho_x$. The probability distribution p plays the role of coding and decoding is a measurement performed on the output quantum system: For each $x \in \mathscr{X}$ a positive operator $F_x \in B(\mathscr{K})$ is given such that $\sum_{x \in \mathscr{X}} F_x \leq I$. (The positive operator $I - \sum_{x \in \mathscr{X}} F_x$ corresponds to failure of decoding.) The output random variable Y is jointly distributed with the input X:

$$\mathrm{Prob}(Y = y | X = x) = \mathrm{Tr}\, \rho_x F_y \qquad (x, y \in \mathscr{X}).$$

and

$$\mathrm{Prob}(Y = y) = \sum_{x \in \mathscr{X}} \mathrm{Prob}(Y = y | X = x)\mathrm{Prob}(X = x) = \mathrm{Tr}\,\rho_X F_y.$$

The probability of error of the coding and decoding is defined as $\mathrm{Prob}(X \neq Y)$.

The channel is used to transfer sequences from the classical alphabet.

A code for the channel $W^{\otimes n}$ is defined by a subset $K_n \subset \mathscr{X}^n$, which is called "codeword set". The codewords are transfered in a memoryless way:

$$W^{\otimes n}(x_1, x_2, \ldots, x_n) = W_{x_1} \otimes W_{x_2} \otimes \ldots \otimes W_{x_n}$$

A number $R > 0$ is called "an ε-achievable rate" for the memoryless channel $\{W^{\otimes n} : n = 1, 2, \ldots\}$ if for every $\delta > 0$ and for sufficiently large n, the channel $\mathscr{E}^{\otimes n}$ has a code with at least $e^{n(R-\delta)}$ codewords and a measurement scheme with error probability not exceeding ε. The largest R which is an ε-achievable rate for every $0 < \varepsilon < 1$ is called "the channel capacity," given by the notation $C_c(W)$.

Theorem 7.8 contains the important result about capacity.

Theorem 7.9. *Let $W : \mathscr{X} \to B(\mathscr{K})$ be a classical-quantum channel. Then*

$$C_c(W) = \sup\{S(W_p) - \sum_{x \in \mathscr{X}} p(x)S(W_x) : p \text{ is a measure on } \mathscr{X}\},$$

where $W_p - \sum_{x \in \mathscr{X}} p(x)W_x$.

7.5 Entanglement-assisted Capacity

The entanglement-assisted capacity of a channel $\mathscr{E}$ is related to a protocol extending the dense coding.

Assume that the bipartite system $\mathscr{H}_A \otimes \mathscr{H}_B$ is in a pure state ρ_{AB}. The classical signal i is encoded by a quantum channel $\mathscr{E}_A^i : B(\mathscr{H}_A) \to B(\mathscr{H}_A)$ and if the signal i is sent, then the receiver receives $(\mathscr{E} \circ \mathscr{E}_A^i \otimes \mathrm{id}_B)(\rho_{AB})$ and he tries to extract the maximal amount of classical information by a measurement on the system B.

The one-shot entanglement-assisted classical capacity is defined as the supremum of some Holevo quantities as

$$C_{ea}^{(1)}(\mathscr{E}) := \sup \left\{ \pi_i S \left(\sum_i (\mathscr{E} \circ \mathscr{E}_A^i \otimes \mathrm{id}_B)(\rho_{AB}) \right) - \sum_i S \left((\mathscr{E} \circ \mathscr{E}_A^i \otimes \mathrm{id}_B)(\rho_{AB}) \right) \right\},$$
$$(7.17)$$

where the supremum is over all possible probability distributions π_i over the signals, all encodings $\mathscr{E}_A^i$ and all pure initial states ρ_{AB}. Using n copies of the channel one has

$$C_{ea}^{(n)}(\mathscr{E}) = C_{ea}^{(1)}(\mathscr{E}^{\otimes n}) \qquad (7.18)$$

and the full entanglement-assisted classical capacity is

$$C_{ea}(\mathscr{E}) = \lim_{n \to \infty} \frac{1}{n} C_{ea}^{(n)}(\mathscr{E}). \qquad (7.19)$$

7.6 Notes

Quantum communication channels were formulated already in the 1970s, see for example the paper [55] summarizing the subject. The capacity problem was unsolved for many years. First Hausladen, Jozsa, Schumacher, Westmoreland and Wooters proved the capacity theorem when all signal states are pure [49]. The result was presented in the *3rd Conference on Quantum Communication and Measurement* in Hakone, Japan, September 1996, by Richard Jozsa. The extension to general signal states, Theorem 7.9, was made by Holevo after the conference in November, but the paper appeared in 1998 [57].

Theorem 7.6 is from [60]. Equivalent forms of the additivity question are discussed by Shor in [110], and for entanglement breaking channels see [109].

7.7 Exercises

1. Let $\mathscr{X} = \mathscr{Y} = \{1, 2, \ldots, n\}$ be the input and output alphabets of a classical channel given by

$$T_{yx} = p\delta(x,y) + \frac{1-p}{n}.$$

Show that the capacity is

$$C = \log n - H \left(p + \frac{1-p}{n}, \frac{1-p}{n}, \ldots, \frac{1-p}{n} \right).$$

2. Compute the Holevo capacity of the phase-dumping channel.
3. Show that the depolarizing channel $\mathscr{E}_{p,2} : M_2(\mathbb{C}) \to M_2(\mathbb{C})$ is not entanglement breaking for $1/2 < p \le 1$.

4. Let $\mathscr{E} : B(\mathscr{H}) \to B(\mathscr{K})$ be a state transformation and $\mathscr{U} \subset B(\mathscr{H})$ be a compact group of unitaries such that (7.11) holds and $\alpha : \mathscr{U} \to B(\mathscr{K})$ is an irreducible group representation. Show that in this case the relative entropy center of the range of $\mathscr{E}$ is the tracial state.

5. Let $\mathscr{E} : M_n(\mathbb{C}) \to M_n(\mathbb{C})$ be an entanglement breaking channel and let $\alpha : M_n(\mathbb{C}) \to M_n(\mathbb{C})$ be a positive trace–preserving mapping. Show that $\alpha \circ \mathscr{E}$ is a state transformation.

Chapter 8
Hypothesis Testing

One of the most basic tasks in quantum statistics is the discrimination of two different quantum states. In the quantum hypothesis testing problem, one has to decide between two states of a system. The state ρ_0 is the **null hypothesis** and ρ_1 is the **alternative hypothesis**. The problem is to decide which hypothesis is true. The decision is performed by a two-valued measurement $\{T, I - T\}$, where $0 \leq T \leq I$ is an observable. T corresponds to the acceptance of ρ_0 and $I - T$ corresponds to the acceptance of ρ_1. T is called a **test**. When the measurement value is 0, the hypothesis ρ_0 is accepted, otherwise the alternative hypothesis ρ_1 is accepted. The quantity $\alpha[T] = \mathrm{Tr}\, \rho_0 (I - T)$ is interpreted as the probability that the null hypothesis is true but the alternative hypothesis is accepted. This is the **error of the first kind**. Similarly, $\beta[T] = \mathrm{Tr}\, \rho_1 T$ is the probability that the alternative hypothesis is true but the null hypothesis is accepted. It is called the **error of the second kind**. Of course, one would like both the first and the second kind of errors to be small.

A single copy of the quantum system is not enough for a good decision. One should make independent measurements on several identical copies, or joint measurements. In the asymptotic theory of the hypothesis testing, the measurements are performed on composite systems. Suppose that a sequence $(\mathscr{H}_n)$ of Hilbert spaces is given, $\rho_0^{(n)}$ and $\rho_1^{(n)}$ are density matrices on $\mathscr{H}_n$. The typical example I have in mind is

$$\rho_0^{(n)} = \rho_0 \otimes \rho_0 \otimes \cdots \otimes \rho_0 \quad \text{and} \quad \rho_1^{(n)} = \rho_1 \otimes \rho_1 \otimes \cdots \otimes \rho_1. \tag{8.1}$$

On the composite system, a positive contraction $T_n \in B(\mathscr{H}_n)$ is considered as a test. So $(T_n, I - T_n)$ represents a $\{0, 1\}$-valued measurement on $\mathscr{H}_n$ for the hypothesis $(\rho_0^{(n)}, \rho_1^{(n)})$. Now the errors of the first and second kind depend on n:

$$\alpha_n[T_n] = \mathrm{Tr}\, \rho_0^{(n)} (I - T_n) \quad \text{and} \quad \beta_n[T_n] = \mathrm{Tr}\, \rho_1^{(n)} T_n. \tag{8.2}$$

In mathematical statistics, a bound is prescribed for the error of the first kind, while the error of the second kind is made small by choosing a large-enough sample size n. This approach leads to the Stein lemma in which the rate of the exponential convergence of the error of the second kind is the relative entropy. In a Bayesian

D. Petz, *Hypothesis Testing*. In: D. Petz, Quantum Information Theory and Quantum Statistics, Theoretical and Mathematical Physics, pp. 109–120 (2008)
DOI 10.1007/978-3-540-74636-2_8   © Springer-Verlag Berlin Heidelberg 2008

setting, a prior probability p is associated to the null hypothesis ρ_0 (and $q := 1 - p$ to the alternative hypothesis). Then the error probability is

$$\text{Err}(T_n) := p\,\alpha[T_n] + q\,\beta[T_n].$$

The aim is the minimalization of the error probability, and the asymptotics is about convergence of the error in the function n. We have again exponential convergence and the rate is the Chernoff bound.

8.1 The Quantum Stein Lemma

Set

$$\beta^*(n,\varepsilon) = \inf\{\text{Tr}\,\rho_1^{(n)}A_n : A_n \in B(\mathscr{H}_n),\quad 0 \le A_n \le I,\quad \text{Tr}\,\rho_0^{(n)}(I - A_n) \le \varepsilon\},$$
$$(8.3)$$

which is the infimum of the error of the second kind when the error of the first kind is at most ε. The importance of this quantity is in the customary approach to hypothesis testing: A bound is prescribed for the error of the first kind, while the error of the second kind is made small by choosing a large-enough sample size n.

The following result is the **quantum Stein lemma**. It is clear from the proof that the asymptotic behavior of $\beta^*(n,\varepsilon)$ remains the same if in the infimum in the definition of $\beta^*(n,\varepsilon)$ is restricted to projection operators A_n.

Theorem 8.1. *In the setting (8.1), the relation*

$$\lim_{n\to\infty}\frac{1}{n}\log\beta^*(n,\varepsilon) = -S(\rho_0\|\rho_1)$$

holds for every $0 < \varepsilon < 1$.

Proof. Let us assume that ρ_0 and ρ_1 are invertible. First we should treat the case when ρ_0 and ρ_1 commute.   □

Let $\delta > 0$ and let $P_n(\delta)$ be the spectral projection of the operator

$$\frac{1}{n}(\log\rho_0^{(n)} - \log\rho_1^{(n)})\tag{8.4}$$

corresponding to the interval $(h - \delta,\ h + \delta)$, where h abbreviates $S(\rho_0\|\rho_1)$. It follows that

$$(h - \delta)P_n(\delta) \le \frac{1}{n}\big(\log\rho_0^{(n)} - \log\rho_1^{(n)}\big)P_n(\delta) \le (h + \delta)P_n(\delta)$$

and hence

$$e^{n(h-\delta)}\rho_1^{(n)}P_n(\delta) \le \rho_0^{(n)}P_n(\delta) \le e^{n(h+\delta)}\rho_1^{(n)}P_n(\delta).\tag{8.5}$$

The expectation value of (8.4) with respect to $\rho_0^{(n)}$ is h and according to the ergodic theorem (which reduces to the law of large numbers in this case), (8.4) converges to h, and we have

$$\lim_{n \to \infty} P_n(\delta) = I$$

(in the strong operator topology in the GNS space of $\rho_0^{(\infty)}$). This implies that $\mathrm{Tr}\,\rho_0^{(n)}\,(I - P_n(\delta)) \to 0$ and for large n, we have $\mathrm{Tr}\,\rho_0^{(n)}\,(I - P_n(\delta)) < \varepsilon$

From the first inequality of (8.5), we can easily conclude

$$\frac{1}{n} \log \mathrm{Tr}\,\rho_1^{(n)} P_n(\delta) \leq -h + \delta.$$

and $\limsup_{n \to \infty} \leq -h$ follows concerning the limit in the statement.

Now let T_n be a positive contraction on $\mathscr{H}_n$ such that $\mathrm{Tr}\,\rho_0^{(n)} T_n \geq 1 - \varepsilon$. This implies

$$\liminf_{n \to \infty} \mathrm{Tr}\,\rho_0^{(n)} P_n(\delta) T_n \geq 1 - \varepsilon$$

since

$$\mathrm{Tr}\,\rho_0^{(n)} T_n P_n(\delta) = \mathrm{Tr}\,\rho_0^{(n)} T_n - \mathrm{Tr}\,\rho_0^{(n)} T_n P_n(\delta)^\perp$$
$$\geq \mathrm{Tr}\,\rho_0^{(n)} T_n - \mathrm{Tr}\,\rho_0^{(n)} P_n(\delta)^\perp.$$

Next we can estimate as follows:

$$\mathrm{Tr}\,\rho_1^{(n)} T_n \geq \mathrm{Tr}\,\rho_1^{(n)} T_n P_n(\delta)$$
$$\geq \mathrm{Tr}\,\rho_0^{(n)} T_n P_n(\delta) e^{-n(h+\delta)}$$
$$= e^{-n(h+\delta)} \cdot \mathrm{Tr}\,\rho_0^{(n)} T_n P_n(\delta)$$

and

$$\frac{1}{n} \log \mathrm{Tr}\,\rho_1^{(n)} T_n \geq -h - \delta + \frac{1}{n} \log \mathrm{Tr}\,\rho_0^{(n)} T_n P_n(\delta).$$

When $n \to \infty$ the last term of the right-hand side converges to 0 and we can conclude that $\liminf_{n \to \infty} \geq -h$ in the statement. The proof is complete under the additional hypothesis that ρ_0 and ρ_1 commute. Although ρ_0 and ρ_1 do not commute in general, the commutator of $n^{-1} \log \rho_0^{(n)}$ and $n^{-1} \log \rho_1^{(n)}$ goes to 0 in norm when $n \to \infty$. Heuristically, this is the reason that the statement holds without the assumption on commutativity.

The next lemma is fundamental; it allows to reduce the general case to the commuting one.

Lemma 8.1. *For $\ell \in \mathbb{N}$, let $\mathscr{C}_\ell$ be the commutant of $\rho_1^{(\ell)}$. Then for any density matrix ω on $\mathscr{H}_\ell$ the inequality*

$$S(\omega\|\rho_1^{(\ell)}) \leq S(\omega_{\mathscr{C}_\ell}\|\rho_1^{(\ell)}) + k\log(\ell+1)$$

holds, where k is the dimension of $\mathscr{H}_1$ and on the right-hand side the reduced density matrix of ω in $\mathscr{C}_\ell$ appears.

Proof. Proof of the lemma. Let

$$\rho_1^{(\ell)} = \sum_{j=1}^{N} \mu_j e_j$$

be the spectral decomposition of the density $\rho_1^{(\ell)}$. N is the number of different eigenvalues; it is easy to see that

$$N \leq (\ell+1)^k.$$

The conditional expectation

$$\mathscr{E}_\ell : a \mapsto \sum_j e_j a e_j \tag{8.6}$$

leaves the state $\rho_1^{(\ell)}$ invariant and maps $B(\mathscr{H}_\ell)$ onto the commutant of $\rho_1^{(\ell)}$. From the conditional expectation property we have

$$S(\omega\|\rho_1^{(\ell)}) = S(\mathscr{E}_\ell(\omega)\|\rho_1^{(\ell)}) + S(\omega\|\mathscr{E}_\ell(\omega)) \tag{8.7}$$

and so $S(\omega\|\mathscr{E}_\ell(\omega))$ is to be estimated for a density ω. It follows from the convexity of $S(\omega\|\mathscr{E}_\ell(\omega))$ that we may assume ω to be pure. Then $S(\omega\|\mathscr{E}_\ell(\omega)) = S(\mathscr{E}_\ell(\omega))$. Each operator $e_j\omega e_j$ has rank 0 or 1. Hence the rank of the density of $\mathscr{E}_\ell(\omega)$ is at most N. It follows that the von Neumann entropy $S(\mathscr{E}_\ell(\omega))$ is majorized by $\log N$. This completes the proof of the lemma. $\qquad\square$

Now let us return to the case when ρ_0 and ρ_1 do not commute. Fix arbitrary $\varepsilon > 0$ and use the lemma to choose ℓ so large that

$$\frac{1}{\ell}S(\rho_0^{(\ell)}\|\rho_1^{(\ell)}) - \frac{1}{\ell}S(\mathscr{E}_\ell(\rho_0^{(\ell)})\|\rho_1^{(\ell)}) \leq \varepsilon, \tag{8.8}$$

where $\mathscr{E}_\ell$ is the conditional expectation onto the subalgebra $\mathscr{C}_\ell$ constructed in the previous lemma. Let $\omega_0 := \mathscr{E}_\ell(\rho_0^{(\ell)})$ and $\omega_1 := \rho_1^{(\ell)}$. These densities commute and the above argument works but needs some completion. The quantity $\beta^*(n,\varepsilon)$ is defined from the tensor powers of ρ_i's but we moved to the powers of ω_i's. The change in the states needs control. This sketches the main idea of the proof of the theorem.

Let $P_{\ell m}(\delta)$ be the spectral projection of

$$\frac{1}{m}(\log \omega_0^{(m)} - \log \omega_1^{(m)}) \qquad (8.9)$$

corresponding to the interval $(S(\omega_0\|\omega_1) - \delta, S(\omega_0\|\omega_1) + \delta)$. Similarly to (8.5), we have

$$e^{m(h(\ell)-\delta)} \omega_1^{(m)} P_{\ell m}(\delta) \leq \omega_0^{(m)} P_{\ell m}(\delta) \leq e^{m(h(\ell)+\delta)} \omega_1^{(m)} P_{\ell m}(\delta), \qquad (8.10)$$

where $h(\ell) = S(\omega_0\|\omega_1)$.

By the law of large numbers

$$\operatorname{Tr} \rho_0^{(m\ell)} P_{\ell m}(\delta) = \operatorname{Tr} \omega_0^{(m)} P_{\ell m}(\delta) \to 1 \qquad (8.11)$$

and

$$\begin{aligned}
\frac{1}{\ell m} \log \operatorname{Tr} \rho_1^{(m\ell)} P_{\ell m}(\delta) &= \frac{1}{\ell m} \log \operatorname{Tr} \omega_1^{(m)} P_{\ell m}(\delta) \\
&\leq \frac{1}{\ell}(-S(\omega_0\|\omega_1) + \delta) + \frac{1}{\ell m} \log \operatorname{Tr} \omega_0^{(m)} P_{\ell m}(\delta) \\
&\leq -S(\rho_0\|\rho_1) + \varepsilon + \frac{\delta}{\ell} + \frac{1}{\ell m} \log \operatorname{Tr} \omega_1^{(m)} P_{\ell m}(\delta).
\end{aligned}$$

Let the test T_n be defined as

$$T_n = P_{\ell m} \quad \text{if} \quad n = m\ell + \ell_0, \quad 0 \leq \ell_0 < \ell.$$

This gives $\limsup_{n\to\infty} \leq -S(\rho_0\|\rho_1)$ in the general case, since ε and δ could be arbitrarily small and the last term goes to 0 as $m \to \infty$. Note that we proved more than it is stated in the theorem because the tests that were proposed are projections.

To prove $\liminf_{n\to\infty} \geq -S(\rho_0\|\rho_1) =: h$, let us take $\delta_1 > 0$ and a sequence T_n of positive contractions such that

$$\operatorname{Tr} \rho_1^{(n)} T_n \leq e^{n(h-\delta_1)}$$

and we shall show that $\operatorname{Tr} \rho_0^{(n)} T_n$ goes to 0.

Choose ℓ such that (8.8) holds. Identify the mth tensor power of $\mathscr{C}_\ell$ with the appropriate commutative subalgebra of $B(\mathscr{H}_{m\ell})$.

Similarly to the definition of $P_{\ell m}(\delta)$, $P_{\ell m}(\delta)'$ can be defined as the spectral projection of

$$-\frac{1}{m} \log \omega_0^{(m)} \qquad (8.12)$$

corresponding to the interval $(S(\omega_0) - \delta, S(\omega_0) + \delta)$. Then

$$e^{m(-S(\omega_0)-\delta)} P_{\ell m}(\delta)' \leq \omega_0^{(m)} P_{\ell m}(\delta)' \leq e^{m(-S(\omega_0)+\delta)} P_{\ell m}(\delta)' \qquad (8.13)$$

and

$$\mathrm{Tr}\,\rho_0^{(m\ell)} P_{\ell m}(\delta)' = \mathrm{Tr}\,\omega_0^{(m)} P_{\ell m}(\delta)' \to 1 \qquad (8.14)$$

Let $Q_{\ell m} = P_{\ell m}(\delta)' \times P_{\ell m}(\delta)$; this is a projection in $B(\mathscr{H}_{m\ell})$.

Lemma 8.2. *Let $Q \le Q_{\ell m}$ be a minimal projection in $B(\mathscr{H}_{\ell m})$. Then we have*

$$\mathrm{Tr}\,\rho_1^{(\ell m)} Q \le \exp\left(m(-S(\omega_0) - h(\ell) + 2\delta)\right) \qquad (8.15)$$

and

$$\mathrm{Tr}\,\rho_1^{(\ell m)} Q \ge \exp\left(m(-S(\omega_0) - h(\ell) - 2\delta)\right) \qquad (8.16)$$

Proof.

$$\begin{aligned}
\mathrm{Tr}\,\rho_1^{(\ell m)} Q &= \mathrm{Tr}\,\omega_1^{(m)} Q \\
&\le \exp\left(m(-S(\omega_0) + \delta)\right) \times \exp\left(-m(h(\ell) - \delta)\right) \times \mathrm{Tr}\,Q \\
&\le \exp\left(m(-S(\omega_0) - h(\ell) + 2\delta)\right)
\end{aligned}$$

from the upper estimates of (8.10) and (8.13). The second inequality is shown similarly but the lower estimates are used. □

Lemma 8.3. *Assume that $\delta_2, \delta_3 > 0$. Then*

$$\mathrm{Tr}\left(Q_{\ell m}\omega_0^{(m)} Q_{\ell m}\right) T_{\ell m} \le \exp\left(-m(S(\omega_0) - \delta_2)\right) \mathrm{Tr}\,Q_{\ell m} T_{\ell m} + \delta_3$$

if m is large.

Proof. Let $E(\ell m, \delta_2)$ be the spectral projection of $Q_{\ell m}\omega_0^{(m)} Q_{\ell m}$ corresponding to the interval $(-\infty, \exp(-mS(\omega_0) - \delta_2))$. Then

$$1 \ge \mathrm{Tr}\,Q_{\ell m}\omega_0^{(m)} Q_{\ell m} \ge \exp\left(-m(S(\omega_0) - \delta_2)\right) \mathrm{Tr}\,E(\ell m, \delta_2)^{\perp}.$$

This is equivalently written as

$$\frac{1}{\ell m} \log \mathrm{Tr}\,E(\ell m, \delta_2)^{\perp} \le S(\omega_0) - \delta_2$$

and implies that

$$\mathrm{Tr}\,\omega_0^{(m)} E(\ell m, \delta_2)^{\perp} \to 0.$$

We have

$$\begin{aligned}
\mathrm{Tr}\left(Q_{\ell m}\omega_0^{(m)} Q_{\ell m}\right) T_{\ell m} &= \mathrm{Tr}\,E(\ell m, \delta_2)\left(Q_{\ell m}\omega_0^{(m)} Q_{\ell m}\right) T_{\ell m} + \mathrm{Tr}\,E(\ell m, \delta_2)^{\perp}\left(Q_{\ell m}\omega_0^{(m)} Q_{\ell m}\right) T_{\ell m} \\
&\le \exp\left(-k(S(\omega_0) - \delta_2)\right) \mathrm{Tr}\,Q_k T_k + \mathrm{Tr}\,\omega_0^{(m)} E(\ell m, \delta_2)^{\perp}
\end{aligned}$$

and for large m the second term is smaller than δ_3. □

Now let us turn to the estimation of $\operatorname{Tr}\rho_0^{(\ell m)}T_{\ell m}$. We have

$$\operatorname{Tr}\rho_0^{(\ell m)}T_{\ell m} \leq 2\operatorname{Tr}\rho_0^{(\ell m)}Q_{\ell m}T_{\ell m}Q_{\ell m} + 2\operatorname{Tr}\rho_0^{(\ell m)}Q_{\ell m}^{\perp}T_{\ell m}Q_{\ell m}^{\perp}. \qquad (8.17)$$

(Recall $T \leq 2QTQ + 2Q^{\perp}TQ^{\perp}$ holds for any positive operator T and for any projection Q, see Exercise 24 in the Appendix). For the second term we have

$$\operatorname{Tr}\rho_0^{(\ell m)}Q_{\ell m}^{\perp}T_{\ell m}Q_{\ell m}^{\perp} \leq \operatorname{Tr}\rho_0^{(\ell m)}Q_{\ell m}^{\perp} \to 0,$$

since in a commutative algebra (8.11) and (8.14) hold.

To handle the first term of (8.17), we can apply the previous lemma. Represent the projector $Q_{\ell m}$ as the sum of minimal projections in $B(\mathscr{H}_{\ell m})$:

$$Q_{\ell m} = \sum_i Q_{\ell m}(i).$$

Since $\exp\left(m(S(\omega_0) + h(\ell) + 2\delta)\right)\operatorname{Tr}\rho_0^{(\ell m)}Q_{\ell m}(i) \geq 1$, we have

$$\begin{aligned}
\operatorname{Tr}Q_{\ell m}T_{\ell m} &= \sum_i \operatorname{Tr}Q_{\ell m}(i)T_{\ell m} \\
&\leq \sum_i \left(\exp\left(m(S(\omega_0) + h(\ell) + 2\delta)\right)\operatorname{Tr}\rho_0^{(\ell m)}Q_{\ell m}(i)\right)\operatorname{Tr}Q_{\ell m}(i)T_{\ell m} \\
&= \exp\left(m(S(\omega_1) + h(\ell) + 2\delta)\right)\operatorname{Tr}\rho_1^{(\ell m)}Q_{\ell m}T_{\ell m}.
\end{aligned}$$

Then

$$\begin{aligned}
\operatorname{Tr}\left(Q_{\ell m}\omega_0^{(m)}Q_{\ell m}\right)T_{\ell m} &\leq \exp\left(-m(S(\omega_0) - \delta_2)\right)\operatorname{Tr}Q_{\ell m}T_{\ell m} + \delta_3 \\
&\leq \exp\left(m(\delta_2 + h(\ell) + 2\delta)\right)\operatorname{Tr}\rho_1^{(\ell m)}Q_{\ell m}T_{\ell m} + \delta_3 \\
&\leq \exp\left(m(\delta_2 + h(\ell) + 2\delta) + \ell m(-S(\rho_0\|\rho_1) - \delta_1)\right) + \delta_3
\end{aligned}$$

The exponent

$$m\ell\left(\frac{\delta_2}{\ell} + \frac{S(\omega_0\|\omega_1)}{\ell} - S(\rho_0\|\rho_1) + \frac{2\delta}{\ell} - \delta_1\right)$$

should be negative to conclude $\operatorname{Tr}\rho_0^{(n)}T_n \to 0$. Recall that δ_1 is given but the other parameters δ, δ_2 and δ_3 can be chosen. The choice

$$\frac{\delta_2}{l} + \frac{2\delta}{\ell} < \delta_1$$

will do. $\square$

8.2 The Quantum Chernoff Bound

The Bayesian error probability is

$$\mathrm{Err}(T_n) = p\mathrm{Tr}\rho_0^{(n)}(I - T_n) + q\mathrm{Tr}\rho_1^{(n)}T_n = p + \mathrm{Tr}\left(T_n(q\rho_1^{(n)} - p\rho_0^{(n)})\right) \qquad (8.18)$$

when the state ρ_0 has prior probability p and $q := 1 - p$. The asymptotics of Chernoff type is about

$$\lim_{n \to \infty} \frac{1}{n} \log\inf\{\mathrm{Err}(T_n) : 0 \le T_n \le I\}.$$

For a fixed n, $\mathrm{Err}(T_n)$ will be minimal if T_n is the support Q_n of the negative part of $q\rho_1^{(n)} - p\rho_0^{(n)}$. The self-adjoint matrix $q\rho_1^{(n)} - p\rho_0^{(n)}$ has some negative eigenvalues and the optimal projection Q_n maps onto the subspace spanned by the corresponding eigenvectors. In general, a self-adjoint matrix A is written as $A = A_+ - A_-$, where the positive matrices A_+ and A_- have orthogonal support and $|A| = A_+ + A_-$. Therefore, $Q_n(q\rho_1^{(n)} - p\rho_0^{(n)}) = -(q\rho_1^{(n)} - p\rho_0^{(n)})_-$ and the minimal error probability is

$$p - \mathrm{Tr}(q\rho_1^{(n)} - p\rho_0^{(n)})_- = \frac{1}{2}\left(1 - \mathrm{Tr}|q\rho_1^{(n)} - p\rho_0^{(n)}|\right) \qquad (8.19)$$

It follows that the Chernoff type asymptotics is

$$\lim_{n \to \infty} \frac{1}{n} \log\left(1 - \frac{\mathrm{Tr}|q\rho_1^{(n)} - p\rho_0^{(n)}|}{2}\right).$$

The limit will be identified by a lower and an upper estimate.

When the densities commute and have eigenvalues $p_0(k)$ and $p_1(k)$, then the minimal error probability is

$$\left(p - \sum_k (qp_1(k) - pp_0(k))_-\right) = \left(\sum_k pp_0(k) - \sum_k (qp_1(k) - pp_0(k))_-\right)$$
$$= \sum_k \min\{pp_0(k), qp_1(k)\}. \qquad (8.20)$$

For the probability distributions p_0 and p_1, we shall use the notation

$$\Delta(p_0, p_1) := \sum_k \min\{pp_0(k), qp_1(k)\}$$

for the minimal error. The classical **Chernoff theorem** is the following.

Theorem 8.2. *In the above notation*

$$\lim_{n \to \infty} \frac{1}{n} \log\Delta(p_0^{n\otimes}, p_1^{n\otimes}) = \inf\left\{\log\sum_k p_0(k)^{1-s}p_1(k)^s : 0 \le s \le 1\right\}$$

holds.

It is remarkable that the effect of the prior probability is washed out in the limit. In the proof of the quantum extension, we shall need the above classical result and the following lemmas.

Lemma 8.4. *Let x, y be two vectors in a Hilbert space $\mathcal{H}$ and λ, μ be positive numbers. Then*

$$\lambda |\langle Px|y\rangle|^2 + \mu |\langle (P^\perp x|y\rangle|^2 \geq \frac{1}{2} |\langle x|y\rangle|^2 \min\{\lambda, \mu\}$$

for any projection P.

Proof. It is sufficient to prove the statement when $\lambda = \mu = 1$. Let $x_1 := Px$ and $x_2 := P^\perp x$. Then $x = x_1 + x_2$. All inner products depend on the the projection of y onto the subspace spanned by x_1 and x_2. Therefore, we may assume that $y = ax_1 + bx_2$. After the reductions our inequality has the form:

$$|a|^2 \|x_1\|^4 + |b|^2 \|x_2\|^4 \geq \frac{1}{2} \Big| a\|x_1\|^2 + b\|x_2\|^2 \Big|^2$$

This obviously holds.   □

Consider two arbitrary density operators ρ_0 and ρ_1 on a k-dimensional Hilbert space $\mathcal{H}$. They have spectral representations

$$\rho_0 = \sum_{i=1}^k \lambda_i |x_i\rangle\langle x_i| \quad \text{and} \quad \rho_1 = \sum_{i=1}^k \mu_i |y_i\rangle\langle y_i|, \tag{8.21}$$

that is, $|x_i\rangle$ and $|y_i\rangle$ $(i = 1, \ldots, k)$ are two orthonormal bases consisting of eigenvectors and λ_i and μ_i are the respective eigenvalues of ρ_0 and ρ_1. Then

$$\beta(\rho_0, \rho_1)_0(i, j) := \lambda_i |\langle x_i|y_j\rangle|^2 \quad \text{and} \quad \beta(\rho_0, \rho_1)_1(i, j) := \mu_j |\langle x_i|y_j\rangle|^2 \tag{8.22}$$

are probability distributions $(1 \leq i, j \leq k)$. They have nice behavior concerning tensor product and Rényi relative entropy.

Lemma 8.5. *The mapping β satisfies the following conditions:*

(i) If ρ_0, ρ_1, ω_0 and ω_1 are density matrices, then

$$\beta(\rho_0, \rho_1)_m \otimes \beta(\omega_0, \omega_1)_m = \beta(\rho_0 \otimes \omega_0, \rho_1 \otimes \omega_1)_m$$

for $m = 0, 1$.
(ii) If u and v are positive numbers and $u + v = 1$, then

$$\text{Tr}\, \rho_0^u \rho_1^v = \sum_{i,j} \beta(\rho_0, \rho_1)_0(i, j)^u \beta(\rho_0, \rho_1)_1(i, j)^v$$

Proof. (i) is an immediate consequence of the definitions. (ii) follows from the formula

$$\mathrm{Tr}\,\rho_0^u \rho_1^v = \sum_i \lambda_i^u \mu_j^v \mathrm{Tr}\,|x_i\rangle\langle x_i||y_j\rangle\langle y_j| = \sum_i \lambda_i^u \mu_j^v |\langle x_i|y_j\rangle|^2$$

□

The Chernoff type lower bound in the asymptotics is due to Nussbaum and Szkola [81]. The very essential point in the proof is the construction (8.22) of probability distributions from density matrices. Given $n \times n$ density matrices, we can obtain probability distributions on an n^2-point-space.

Theorem 8.3. *Let ρ_0 and ρ_1 be density matrices and $(T_n, I - T_n)$ be a sequence of tests to distinguish them. The Bayesian error probability (8.18) has the asymptotics*

$$\liminf_{n \to \infty} \frac{1}{n} \log \mathrm{Err}(T_n) \geq \inf\{\log \mathrm{Tr}\,\rho_0^{1-s} \rho_1^s : 0 \leq s \leq 1\}.$$

Proof. The optimal tests are projections, so in place of T_n we can put projections P_n:

$$\mathrm{Err}(P_n) = p\mathrm{Tr}\,P_n^\perp \rho_0^{(n)} + q\mathrm{Tr}\,P_n \rho_1^{(n)}$$

Now we can use the spectral decompositions of $\rho_0^{(n)}$ and $\rho_1^{(n)}$, similarly to (8.21), and have

$$\mathrm{Err}(P_n) = p\mathrm{Tr}\,P_n^\perp \rho_0^{(n)} + q\mathrm{Tr}\,\rho_1^{(n)} P_n$$

$$= \sum_{i,j} \left(p\lambda_i(n)|\langle P_n^\perp x_i(n)|y_j(n)\rangle|^2 + q\mu_j(n)|\langle P_n x_i(n)|y_j(n)\rangle|^2 \right),$$

since

$$\mathrm{Tr}\,P_n^\perp |x_i(n)\rangle\langle x_i(n)| = \|P_n^\perp x_i(n)\|^2 = \sum_j |\langle P_n x_i(n)|y_j(n)\rangle|^2$$

and similarly for the other term. Now we can use Lemma 8.4 to arrive at a lower bound:

$$\mathrm{Err}(P_n) \geq \frac{1}{2} \sum_{i,j} \min\{p\lambda_i(n), q\mu_j(n)\}|\langle x_i(n)|y_j(n)\rangle|^2$$

The right-hand side is

$$\frac{1}{2}\Delta(\beta(\rho_0^{n\otimes}, \rho_1^{n\otimes})_0, \beta(\rho_0^{n\otimes}, \rho_1^{n\otimes})_1) = \frac{1}{2}\Delta(\beta(\rho_0, \rho_1))_0^{n\otimes}, \beta(\rho_0, \rho_1)_1^{n\otimes})$$

and we have

$$\liminf_{n\to\infty} \frac{1}{n}\log \text{Err}(P_n) \geq \liminf_{n\to\infty} \frac{1}{n}\Delta(\beta(\rho_0,\rho_1)_0^{n\otimes}, \beta(\rho_0,\rho_1)_1^{n\otimes})$$

$$= \inf\left\{ \log\sum_{i,j}\beta(\rho_0,\rho_1)_0(i,j)^{1-s}\beta(\rho_0,\rho_1)_1(i,j)^s : 0\leq s\leq 1\right\}$$

$$= \inf\left\{ \log\text{Tr}\,\rho_0^{1-s}\rho_1^s : 0\leq s\leq 1\right\},$$

where the classical Chernoff theorem was used. $\square$

The upper estimate for the Chernoff type asymptotics is due to Audenaert et al. [10].

Theorem 8.4. *Let ρ_0 and ρ_1 be density matrices and let Q_n be the support of the negative part of $\rho_1^{(n)} - \rho_0^{(n)}$ to be used for testing. The Bayesian error probability (8.18) has the asymptotics*

$$\limsup_{n\to\infty} \frac{1}{n}\log\text{Err}(Q_n) \leq \inf\{\log\text{Tr}\,\rho_0^{1-s}\rho_1^s : 0\leq s\leq 1\}.$$

Proof. We have to estimate

$$\limsup_{n\to\infty} \frac{1}{n}\log(1 - \text{Tr}\,|p\rho_0^{(n)} - q\rho_1^{(n)}|) \qquad (8.23)$$

according to (8.19). Since the trace inequality

$$\frac{1}{2}\text{Tr}\,(A+B-|A-B|) \leq \text{Tr}\,A^s B^{1-s} \qquad (8.24)$$

holds for positive operators A, B and for any number $0\leq s\leq 1$ (see the Appendix), we have

$$1 - \text{Tr}\,|p\rho_0^{(n)} - q\rho_1^{(n)}| \leq 2q^s p^{1-s}\text{Tr}\,(\rho_1^{(n)})^s(\rho_0^{(n)})^{1-s} = 2q^s p^{1-s}(\text{Tr}\,\rho_1^s\rho_0^{1-s})^n.$$

This gives the upper bound $\log\text{Tr}\,\rho_0^{1-s}\rho_1^s$ for (8.23). $\square$

8.3 Notes

The quantum Stein lemma was conjectured by Hiai and Petz in [53], where $\limsup \leq -S(\rho_0\|\rho_1)$ was also proved. The converse inequality was first shown by Ogawa and Nagaoka in the general case [82]. The proof presented here benefits from the dissertation [15] (see also [17]).

The lower estimate for the quantum Chernoff bound is due to Nussbaum and Szkola [81] and the upper estimate was proven by Audenaert et al. [10], both results are from 2006. The proof of the classical Chernoff theorem, Theorem 8.2 is available in the book [26].

8.4 Exercises

1. Compute the marginals of the probability distributions $\beta(\rho_0, \rho_1)_0$ and $\beta(\rho_0, \rho_1)_1$ in (8.22).

2. Let $f(s) = \mathrm{Tr}\,\rho_1^s \rho_0^{1-s}$. Show that f is convex. What are $f'(0)$ and $f'(1)$?

Chapter 9
Coarse-grainings

A quantum mechanical system is described by an algebra $\mathcal{M}$ of operators, and the dynamical variables (or observables) correspond to the self-adjoint elements. The evolution of the system $\mathcal{M}$ can be described in the **Heisenberg picture**, in which an observable $A \in \mathcal{M}$ moves into $\alpha(A)$, where α is a linear transformation. α is induced by a unitary in case of the time evolution of a closed system but it could be the irreversible evolution of an open system. The **Schrödinger picture** is dual, it gives the transformation of the states:

$$\langle \alpha(A), \rho \rangle = \langle A, \mathcal{E}(\rho) \rangle,$$

where $\mathcal{E}$ is the state transformation and the duality means $\langle B, \omega \rangle = \mathrm{Tr}\, \omega B$. State transformation is an essential concept in quantum information theory and its role is the performance of information transfer. However, state transformation may appear in a different context as well. Shortly speaking, the dual of a state transformation will be called **coarse-graining**, in particular in a statistical context.

Assume that a level of observation does not allow to know the expectation value of all observables, but only a part of them. In our algebraic approach it is assume that this part is the self-adjoint subalgebra $\mathcal{N}$ of the full algebra $\mathcal{M}$. The positive linear embedding $\alpha : \mathcal{N} \to \mathcal{M}$ is a coarse-graining. It provides partial information of the total quantum system $\mathcal{M}$. If the algebra $\mathcal{N}$ is "small" compared with $\mathcal{M}$, then loss of information takes place and the problem of statistical inference is to reconstruct the real state of $\mathcal{M}$ from partial information.

9.1 Basic Examples

Assume that the Hilbert space describing our quantum system is $\mathcal{H}$. A completely positive identity preserving linear mapping from an algebra $\mathcal{N}$ to $B(\mathcal{H})$ will be called **coarse-graining**. It will be mostly assumed that $\mathcal{N}$ is a subalgebra of $B(\mathcal{K})$, where $\mathcal{K}$ is another Hilbert space. In the algebraic approach followed here, this

D. Petz, *Coarse-grainings*. In: D. Petz, Quantum Information Theory and Quantum Statistics, Theoretical and Mathematical Physics, pp. 121–142 (2008)
DOI 10.1007/978-3-540-74636-2_9

Hilbert space is not specified always. It is well known that a completely unit-preserving mapping α satisfies the **Schwarz inequality**

$$\alpha(A^*A) \geq \alpha(A)^*\alpha(A). \tag{9.1}$$

Example 9.1. The simplest example of coarse-graining appears if a component of a composite system $\mathcal{H} \equiv \mathcal{H}_1 \otimes \mathcal{H}_2$ is neglected. Assume that we restrict ourselves to the observables of the first subsystem. Then the embedding

$$\alpha : B(\mathcal{H}_1) \to B(\mathcal{H}), \qquad \alpha : A \mapsto A \otimes I_2$$

is the relevant coarse-graining. This is not only positive, but also a multiplicative isometry; therefore it is called "embedding."

The dual of α is the **partial trace** Tr_2:

$$\langle \alpha(A), B \otimes C \rangle = \mathrm{Tr}\,(A^*B) \otimes C = \mathrm{Tr}\,A^*B\,\mathrm{Tr}\,C = \langle A, \mathcal{E}(B \otimes C) \rangle,$$

where $\mathcal{E}$ defined as $\mathcal{E}(B \otimes C) = B\,\mathrm{Tr}\,C$ is the so-called partial trace over the second factor. $\square$

Example 9.2. Consider a composite system of n identical particles: $\mathcal{H} \otimes \mathcal{H} \otimes \ldots \otimes \mathcal{H}$ and assume that we restrict ourselves to the symmetric observables. Each permutation of the particles induces a unitary U and set $\mathcal{N} := \{A : UAU^* = A$ for every permutation unitary $U\}$. The embedding of $\mathcal{N}$ is again a coarse-graining. The algebra $\mathcal{N}$ is not isomorphic to a full matrix algebra but it is a subalgebra of $B(\mathcal{H}^{n\otimes})$. $\square$

The coarse-graining of the previous two examples were embeddings which are multiplicative. A general coarse-graining is not fully multiplicative but it satisfies a restricted multiplicativity.

Lemma 9.1. *Let $\alpha : \mathcal{N} \to \mathcal{M}$ be a coarse-graining. Then*

$$\mathcal{N}_\alpha := \{A \in \mathcal{N} : \alpha(A^*A) = \alpha(A)^*\alpha(A) \text{ and } \alpha(AA^*) = \alpha(A)\alpha(A)^*\} \tag{9.2}$$

is a subalgebra of $\mathcal{N}$ and

$$\alpha(AB) = \alpha(A)\alpha(B) \quad \text{and} \quad \alpha(BA) = \alpha(B)\alpha(A) \tag{9.3}$$

hold for all $A \in \mathcal{N}_\alpha$ and $B \in \mathcal{N}$.

Proof. The proof is based only on the Schwarz inequality (9.1). Assume that $\alpha(AA^*) = \alpha(A)\alpha(A)^*$. Then

$$\begin{aligned}
t\big(\alpha(A)\alpha(B) + \alpha(B)^*\alpha(A)^*\big) &= \alpha(tA^*+B)^*\alpha(tA^*+B) - t^2\alpha(A)\alpha(A)^* - \alpha(B)^*\alpha(B) \\
&\leq \alpha\big((tA^*+B)^*(tA^*+B)\big) - t^2\alpha(AA^*) - \alpha(B)^*\alpha(B) \\
&= t\alpha(AB + B^*A^*) + \alpha(B^*B) - \alpha(B)^*\alpha(B)
\end{aligned}$$

for a real t. Divide the inequality by t and let $t \to \pm\infty$. Then

$$\alpha(A)\alpha(B) + \alpha(B)^*\alpha(A)^* = \alpha(AB + B^*A^*)$$

and similarly

$$\alpha(A)\alpha(B) - \alpha(B)^*\alpha(A)^* = \alpha(AB - B^*A^*).$$

Adding these two equalities we have

$$\alpha(AB) = \alpha(A)\alpha(B).$$

The other identity is proven similarly.   □

The subalgebra $\mathcal{N}_\alpha$ is called the **multiplicative domain** of α.

Example 9.3. The algebra of a quantum system is typically non-commutative but the mathematical formalism supports commutative algebras as well. A measurement is usually modeled by a positive partition of unity $(F_i)_{i=1}^n$, where F_i is a positive operator in $\mathcal{M} = B(\mathcal{H})$ and $\sum_i F_i = I$. The mapping $\beta : \mathbb{C}^n \to \mathcal{M}$, $(z_1, z_2, \ldots, z_n) \mapsto \sum_i z_i F_i$ is positive and unital mapping from the commutative algebra $\mathbb{C}^n$ to the non-commutative algebra $\mathcal{M}$. Every positive unital mappings $\mathbb{C}^n \to \mathcal{M}$ corresponds in this way to a certain measurement. Measurements are coarse-grainings from commutative algebras.

The n-tuple $(z_1, z_2, \ldots, z_n)$ can be regarded as a diagonal matrix and $\mathcal{N}$ is a subalgebra of a matrix algebra.   □

Assume that $\mathcal{A}$ is a subalgebra of $B(\mathcal{H})$. If ρ is a density matrix in $B(\mathcal{H})$, then the **reduced density** matrix $\rho_0 \in \mathcal{A}$ is uniquely determined by the equation

$$\mathrm{Tr}\,\rho_0 A = \mathrm{Tr}\,\rho A \quad \text{for every } A \in \mathcal{A}.$$

The mapping $\mathcal{E} : \rho \mapsto \rho_0$ is a state transformation and it is a sort of generalization of the partial trace. Sometimes $\rho|\mathcal{A}$ is written instead of ρ_0. For example,

$$S(\rho|\mathcal{A} \,\|\, \omega|\mathcal{A}) := S(\rho_0 \| \omega_0)$$

if ρ and ω are density matrices.

9.2 Conditional Expectations

Conditional expectation is a transformation of the observables. Let $\mathcal{A}$ be a subalgebra of $B(\mathcal{H})$ such that $\mathcal{A}^* = \mathcal{A}$ and $I_\mathcal{H} \in \mathcal{A}$. More generally, in place of $B(\mathcal{H})$ we can consider a matrix algebra $\mathcal{B}$. A **conditional expectation** $E : \mathcal{B} \to \mathcal{A}$ is a unital positive mapping which has the property

$$E(AB) = AE(B) \quad \text{for every } \quad A \in \mathcal{A} \quad \text{and} \quad B \in \mathcal{B}. \tag{9.4}$$

Choosing $B = I$, it can be seen that E acts identically on $\mathscr{A}$. It follows from the positivity of E that $E(B^*) = E(B)^*$. Therefore,

$$E(BA) = E((A^*B^*)^*) = E(A^*B^*)^* = (A^*E(B^*))^* = E(B^*)^*A = E(B)A \qquad (9.5)$$

for every $A \in \mathscr{A}$ and $B \in \mathscr{B}$.

Example 9.4. Let $\mathscr{B}$ be a matrix algebra and $\rho \in \mathscr{B}$ an invertible density matrix. Assume that $\alpha : \mathscr{B} \to \mathscr{B}$ is a coarse-graining such that

$$\operatorname{Tr}\rho\,\alpha(B) = \operatorname{Tr}\rho B \quad \text{for every } B \in \mathscr{B}.$$

Let $\mathscr{A} := \{A \in \mathscr{B} : \alpha(A) = A\}$ be the set of fixed points of α. First I show that $\mathscr{A}$ is an algebra. Assume that $A \in \mathscr{A}$. Then

$$\operatorname{Tr}\rho A^*A = \operatorname{Tr}\rho\,\alpha(A^*A) \geq \operatorname{Tr}\rho\,\alpha(A)^*\alpha(A) = \operatorname{Tr}\rho A^*A.$$

Since ρ is invertible, $\alpha(A^*A) = A^*A$ and $A^*A \in \mathscr{A}$ and similarly $AA^* \in \mathscr{A}$. This gives that $\mathscr{A} \subset \mathscr{N}_\alpha$. Therefore for $A_1, A_2 \in \mathscr{A}$ we have $\alpha(A_1A_2) = \alpha(A_1)\alpha(A_2) = A_1A_2$ and $A_1A_2 \in \mathscr{A}$. Hence we can conclude that $\mathscr{A}$ is a subalgebra of $\mathscr{N}_\alpha$.

When $\mathscr{B}$ is endowed with the inner product $\langle B_1, B_2 \rangle := \operatorname{Tr}\rho B_1^*B_2$, then α becomes a contraction and as per the **von Neumann ergodic theorem**

$$s_n(B) := \frac{1}{n}(B + \alpha(B) + \ldots + \alpha^{n-1}(B)) \to E(B),$$

where $E : \mathscr{B} \to \mathscr{A}$ is an orthogonal projection. Since s_n's are coarse-grainings, so is their limit E. In fact, E is a conditional expectation onto the fixed-point algebra $\mathscr{A}$. It is easy to see that for $A \in \mathscr{A}$ and $B \in \mathscr{B}$, $s_n(AB) = As_n(B)$. The limit $n \to \infty$ gives $E(AB) = AE(B)$.

A conditional expectation $E : \mathscr{B} \to \mathscr{A}$ was obtained as the limit of s_n under the condition that there is an invertible state left by α invariant. (This result is often referred to as the **Kovács–Szűcs theorem**.) $\square$

Heuristically, $E(B)$ is a kind of best approximation of B from $\mathscr{A}$. This is justified in the next example.

Example 9.5. Assume that τ is a linear functional on a matrix algebra $\mathscr{B}$ such that

(i) $\tau(B) \geq 0$ if $B \geq 0$,
(ii) If $B \geq 0$ and $\tau(B) = 0$, then $B = 0$,
(iii) $\tau(B_1B_2) = \tau(B_2B_1)$.

(These conditions say that τ is a positive, faithful and tracial functional.) $\mathscr{B}$ becomes a Hilbert space when it is endowed with the inner product

$$\langle B_1, B_2 \rangle := \tau(B_1^*B_2).$$

Recall that $B \in \mathscr{B}$ is positive if (and only if) $\tau(BB_1) \geq 0$ for every positive B_1.

Let $\mathscr{A}$ be a unital subalgebra of $\mathscr{B}$. We can claim that the orthogonal projection E with respect to the above-defined inner product is a conditional expectation of $\mathscr{B}$ onto $\mathscr{A}$.

Since $I \in \mathscr{A}$, we have $E(I) = I$ and E is unital. Let $B \in \mathscr{B}$ be positive. To show the positivity of E, we have to show that $E(B) \geq 0$. We have

$$\tau(A_0 E(B)) = \langle A_0^*, E(B) \rangle = \langle E(A_0^*), B \rangle = \langle A_0^*, B \rangle = \tau(A_0 B) \geq 0$$

for every positive $A_0 \in \mathscr{A}$. It follows that $E(B) \in \mathscr{A}$ is positive.

Condition (9.4) is equivalent to

$$\tau(A_1 E(AB)) = \tau(A_1 A E(B))$$

for every $A_1 \in \mathscr{A}$. This is true since

$$\tau(A_1 E(AB)) = \langle A_1^* A, E(AB) \rangle = \langle E(A_1^*), AB \rangle = \langle A_1^*, AB \rangle = \tau(A_1 AB)$$

and

$$\tau(A_1 A E(B)) = \langle A^* A_1^*, E(B) \rangle = \langle E(A^* A_1^*), B \rangle = \langle A^* A_1^*, B \rangle = \tau(A_1 AB).$$

In the proof of both the positivity and the module property (9.4) of E the tracial condition (iii) of τ was used.   □

A conditional expectation $E : \mathscr{B} \to \mathscr{A}$ is automatically **completely positive**. For $A_i \in \mathscr{A}$ and $B_i \in \mathscr{B}$ we have

$$\sum_{ij} A_i^* E(B_i^* B_j) A_j = E\left(\left(\textstyle\sum_i B_i A_i \right)^* \left(\textstyle\sum_j B_j A_j \right) \right) \geq 0 \qquad (9.6)$$

due to the positivity and the module property of E.

The mapping $\mathscr{E}$ in Example 3.6 is a conditional expectation from the $n \times n$ matrices to the algebra of diagonal $n \times n$ matrices.

Given a conditional expectation $E : B(\mathscr{H}) \to \mathscr{A}$ and a density matrix $\rho_0 \in \mathscr{A}$, the formula

$$\operatorname{Tr} \rho B = \operatorname{Tr} \rho_0 E(B) \qquad (B \in B(\mathscr{H}))$$

determines a density ρ such that the reduced density is ρ_0. The correspondence $\mathscr{E} : \rho_0 \mapsto \rho$ is a state transformation and called **state extension**. The state extension $\mathscr{E}$ is the dual of the conditional expectation E. The converse is also true.

Theorem 9.1. *Let $\mathscr{A}$ be a subalgebra of the matrix algebra $\mathscr{B}$. If $\mathscr{E} : \mathscr{A} \to \mathscr{B}$ is a positive trace-preserving mapping such that the reduced state of $\mathscr{E}(\rho_0)$ is ρ_0 for every density $\rho_0 \in \mathscr{A}$, then the dual of $\mathscr{E}$ is a conditional expectation.*

Proof. The dual $E : \mathscr{B} \to \mathscr{A}$ is a positive unital mapping and $E(A) = A$ for every $A \in \mathscr{A}$. For a contraction B, $\|E(B)\|^2 = \|E(B)^* E(B)\| \leq \|E(B^* B)\| \leq \|E(I)\| = 1$. Therefore, we have $\|E\| = 1$.

Let P be a projection in $\mathscr{A}$ and $B_1, B_2 \in \mathscr{B}$. We have

$$
\begin{aligned}
\|PB_1 + P^\perp B_2\|^2 &= \|(PB_1 + P^\perp B_2)^*(PB_1 + P^\perp B_2)\| \\
&= \|B_1^* PB_1 + B_2^* P^\perp B_2\| \\
&\leq \|B_1^* PB_1\| + \|B_2^* P^\perp B_2\| \\
&= \|PB_1\|^2 + \|P^\perp B_2\|^2.
\end{aligned}
$$

Using this, we estimate for an arbitrary $t \in \mathbb{R}$ as follows.

$$
\begin{aligned}
(t+1)^2 \|P^\perp E(PB)\|^2 &= \|P^\perp E(PB) + tP^\perp E(PB))\|^2 \\
&\leq \|PB + tP^\perp E(PB)\|^2 \\
&\leq \|PB\|^2 + t^2 \|P^\perp E(PB)\|^2
\end{aligned}
$$

Since t can be arbitrary, $P^\perp E(PB) = 0$, that is, $PE(PB) = E(PB)$. We may write $P^\perp$ in place of P:

$$
(I - P)E((I - P)B) = E((I - P)B), \text{ equivalently, } PE(B) = PE(PB).
$$

Therefore we can conclude $PE(B) = E(PB)$. The linear span of projections is the full algebra $\mathscr{A}$ and we have $AE(B) = E(AB)$ for every $A \in \mathscr{A}$. This completes the proof.   □

It is remarkable that in the proof of the previous theorem it was shown that if $E : \mathscr{B} \to \mathscr{A}$ is a positive mapping and $E(A) = A$ for every $A \in \mathscr{A}$, then E is a conditional expectation. This statement is called **Tomiyama theorem**. Actually, the positivity is equivalent to the condition $\|E\| = 1$ (when $E(I) = I$ is assumed). This has the consequence that in the definition of the conditional expectation, it is enough if (9.4) holds for $B = I$.

We can say that the conditional expectation $E : B(\mathscr{H}) \to \mathscr{A}$ preserves the state ρ if

$$
\operatorname{Tr} \rho B = \operatorname{Tr} \rho E(B) \quad \text{for every} \quad B \in B(\mathscr{H}). \tag{9.7}
$$

Takesaki's theorem tells about the existence of a conditional expectation.

Theorem 9.2. *Let $\mathscr{B} \simeq M_n(\mathbb{C})$ be a matrix algebra and $\mathscr{A}$ be its subalgebra. Suppose that $\rho \in \mathscr{B}$ is an invertible density matrix. The following conditions are equivalent:*

(i) A conditional expectation $E : \mathscr{B} \to \mathscr{A}$ preserving ρ exists.
(ii) For every $A \in \mathscr{A}$ and for the reduced density $\rho_0 \in \mathscr{A}$

$$
\rho^{1/2} A \rho^{-1/2} = \rho_0^{1/2} A \rho_0^{-1/2} \tag{9.8}
$$

holds.
(iii) For every $A \in \mathscr{A}$

$$
\rho^{1/2} A \rho^{-1/2} \in \mathscr{A} \tag{9.9}
$$

holds.

Proof. Recall that the reduced density in $\mathscr{A}$ is determined by the equation

$$\mathrm{Tr}\,\rho A = \mathrm{Tr}\,\rho_0 A \quad \text{for every} \quad A \in \mathscr{A}.$$

Assume that $E : \mathscr{B} \to \mathscr{A}$ is a conditional expectation preserving ρ. We can consider $\mathscr{B}$ as a Hilbert space with the inner product

$$\langle B_1, B_2 \rangle = \mathrm{Tr}\,\rho B_1^* B_2.$$

Then the adjoint of the embedding $\mathscr{A} \to \mathscr{B}$ is the conditional expectation $E : \mathscr{B} \to \mathscr{A}$:

$$\langle A, B \rangle = \mathrm{Tr}\,\rho A^* B = \mathrm{Tr}\,\rho E(A^* B) = \mathrm{Tr}\,\rho A^* E(B) = \langle A, E(B) \rangle$$

for $A \in \mathscr{A}$ and $B \in \mathscr{B}$.

Define a conjugate linear operator:

$$S : \mathscr{B} \to \mathscr{B}, \quad S(B) = B^* \quad (B \in \mathscr{B}).$$

We can compute its adjoint S^*, which is determined by the equation

$$\langle S(B_1), B_2 \rangle = \langle S^*(B_2), B_1 \rangle \quad (B_1, B_2 \in \mathscr{B}).$$

We can show that $S^*(B_2) = \rho B_2^* \rho^{-1}$:

$$\langle S(B_1), B_2 \rangle = \mathrm{Tr}\,\rho B_1 B_2 = \overline{\mathrm{Tr}\,B_2^* B_1^* \rho} = \overline{\mathrm{Tr}\,\rho B_1^* \rho B_2^* \rho^{-1}} = \overline{\langle B_1, \rho B_2^* \rho^{-1} \rangle}.$$

Due to the positivity of the conditional expectation, $ES = SE$. This implies that the positive operator $\Delta := S^* S$ leaves the subspace $\mathscr{A}$ invariant. The action of Δ is

$$\Delta B = \rho B \rho^{-1} \quad (B \in \mathscr{B}).$$

For $A \in \mathscr{A}$, we have

$$\langle A, \Delta A \rangle = \mathrm{Tr}\,\rho A A^* = \mathrm{Tr}\,\rho_0 A A^* = \langle A, \Delta_0 A \rangle$$

if $\Delta_0 A = \rho_0 A \rho_0^{-1}$. Since the restriction of Δ to $\mathscr{A}$ is Δ_0, we have

$$\Delta_0^{1/2} A = \rho_0^{1/2} A \rho_0^{-1/2} = \Delta^{1/2} A = \rho^{1/2} A \rho^{-1/2} \quad (A \in \mathscr{A}).$$

This is exactly (9.8) and (i) $\Rightarrow$ (ii) is proven.

The conditional expectation $F : \mathscr{B} \to \mathscr{A}$ preserving Tr exists; it is constructed in Example 9.5. The mapping

$$E_\rho(B) := \rho_0^{-1/2} F(\rho^{1/2} B \rho^{1/2}) \rho_0^{-1/2} \tag{9.10}$$

is completely positive and preserves the state ρ. Indeed,

$$\mathrm{Tr}\,\rho E_\rho(B) = \mathrm{Tr}\,\rho_0 E_\rho(B) = \mathrm{Tr}\,F(\rho^{1/2} B \rho^{1/2}) = \mathrm{Tr}\,(\rho^{1/2} B \rho^{1/2}) = \mathrm{Tr}\,\rho B.$$

E_ρ is canonically determined and it is often called **generalized conditional expectation**.

Under Takesaki's condition (9.8), it can be shown that E_ρ is really a conditional expectation. Actually, we can prove more. Assume that $A \in \mathscr{A}$ satisfies the condition $\rho^{1/2}A\rho^{-1/2} = \rho_0^{1/2}A\rho_0^{-1/2}$. Then

$$E_\rho(A) = \rho_0^{-1/2}F(\rho^{1/2}A\rho^{-1/2}\rho)\rho_0^{-1/2} = \rho_0^{-1/2}F(\rho_0^{1/2}A\rho_0^{-1/2}\rho)\rho_0^{-1/2}$$
$$= \rho_0^{-1/2}\rho_0^{1/2}A\rho_0^{-1/2}F(\rho)\rho_0^{-1/2} = A\rho_0^{-1/2}F(\rho)\rho_0^{-1/2} = A.$$

It follows that $A \in \mathscr{A}$ is in the multiplicative domain of E_ρ. If this holds for every $A \in \mathscr{A}$, then E is really a conditional expectation. This completes the proof of (ii) $\Rightarrow$ (i).

(ii) $\Rightarrow$ (iii) is obvious and we can show the converse. Endow $\mathscr{B}$ with the Hilbert–Schmidt inner product. Condition (iii) tells us that the positive operator

$$\Delta B = \rho B \rho^{-1} \qquad (B \in \mathscr{B})$$

leaves the subspace $\mathscr{A}$ invariant. This is true also for Δ^{it}. For every $A \in \mathscr{A}$,

$$\rho^{it}A\rho^{-it} \in \mathscr{A}.$$

Now let us apply Theorem 11.26 for the unitaries ρ^{it}. In the decomposition of $\mathscr{A}$

$$P_{(m,d)}\rho = \sum_{i=1}^{K(m,d)} \rho_i^L \otimes \rho_i^R \otimes E_{i,i}^{m,d},$$

since the permutation σ must be identity. We have

$$P_{(m,d)}\rho_0 = \sum_{i=1}^{K(m,d)} I_m \otimes \rho_i^R \otimes E_{i,i}^{m,d}.$$

Assume that $A \in \mathscr{A}$. Then $P_{(m,d)}A$ has the form

$$\sum_{i=1}^{K(m,d)} I_m \otimes A(m,d,i) \otimes E_{i,i}^{m,d}.$$

and

$$P_{(m,d)}\rho^{1/2}A\rho^{-1/2} = \sum_{i=1}^{K(m,d)} I_m \otimes (\rho_i^R)^{1/2}A(m,d,i)(\rho_i^R)^{-1/2} \otimes E_{i,i}^{m,d}$$

and this is the same as $P_{(m,d)}\rho_0^{1/2}A\rho_0^{-1/2}$. Hence (iii) $\Rightarrow$ (ii) is shown. □

An important property of the relative entropy is related to conditional expectation.

Theorem 9.3. *Let $\mathscr{A} \subset B(\mathscr{H})$ be a subalgebra and $\rho, \omega \in B(\mathscr{H})$ be the density matrices with reductions $\rho_0, \omega_0 \in \mathscr{A}$. Assume that there exists a conditional expec-*

tation $E : B(\mathcal{H}) \to \mathcal{A}$ *onto* $\mathcal{A}$ *which leaves* ω *invariant. Then*

$$S(\rho\|\omega) = S(\rho_0\|\omega_0) + S(\rho\|\rho \circ E). \tag{9.11}$$

Proof. Denoting the density of $\rho \circ E$ by ρ_1, we need to show that

$$\mathrm{Tr}\,\rho \log \rho - \mathrm{Tr}\,\rho \log \omega = \mathrm{Tr}\,\rho \log \rho_0 - \mathrm{Tr}\,\rho \log \omega_0 + \mathrm{Tr}\,\rho \log \rho - \mathrm{Tr}\,\rho \log \rho_1.$$

The conditional expectation E leaves the states ω and $\rho \circ E$ invariant. According to (iii) in Theorem 9.8 this condition implies

$$\rho_1^{it}\omega^{-it} = \rho_0^{it}\omega_0^{-it} \qquad (t \in \mathbb{R})$$

Differentiating this at $t = 0$, we have

$$\log \rho_1 - \log \omega = \log \rho_0 - \log \omega_0.$$

This relation gives the proof. $\qquad\qquad\qquad\qquad\qquad\qquad\qquad\qquad\qquad\qquad\square$

The state $\rho \circ E$ is nothing else than the extension of $\rho_0 \in \mathcal{A}$ with respect to the state ω. Therefore, the **conditional expectation property** (9.11) has the following heuristical interpretation. The relative entropy distance of ρ and ω comes from two sources. First, the distance of their restrictions to $\mathcal{A}$; second, the distance of ρ from the extension of ρ_0 (with respect to ω).

Note that the conditional expectation property is equivalently written as

$$S(\rho\|\omega) = S(\rho\|\rho \circ E) + S(\rho \circ E\|\omega). \tag{9.12}$$

Theorem 9.4. *Let* $\mathcal{B}$ *be a matrix algebra and* $\mathcal{A}$ *be its subalgebra. Suppose that* $\omega, \rho \in \mathcal{B}$ *are invertible density matrices. Then* $E_\rho = E_\omega$ *if and only if*

$$\rho_0^{-1/2}\rho^{1/2} = \omega_0^{-1/2}\omega^{1/2}$$

holds for the reduced densities $\omega_0, \rho_0 \in \mathcal{A}$.

Proof. Since

$$E_\rho(B) = F(\rho_0^{-1/2}\rho^{1/2}B\rho^{1/2}\rho_0^{-1/2}),$$

the condition obviously imply $E_\rho = E_\omega$. To see the converse, refer Theorem 9.7 and it can be noted that

$$\rho_0^{-1/2}\omega_0^{1/2} = \rho^{-1/2}\omega^{1/2}$$

is an equivalent form of the condition. $\qquad\qquad\qquad\qquad\qquad\qquad\qquad\qquad\square$

A conditional expectation preserving a given state ρ does not always exist. This simple fact has very far reaching consequences. Those concepts and arguments in classical probability which are based on conditioning have a very restricted chance to be extended to the quantum setting.

Example 9.6. Let ρ_{12} be a state of the composite system $\mathcal{H}_1 \otimes \mathcal{H}_2$. Then a conditional expectation $B(\mathcal{H}_1 \otimes \mathcal{H}_2) \rightarrow B(\mathcal{H}_1) \otimes \mathbb{C}I_2$ (preserving ρ_{12}) exists if and only if ρ_{12} is a product state.

When $\rho_{12} = \rho_1 \otimes \rho_2$, then

$$E(X \otimes Y) = X \operatorname{Tr} \rho_2 Y \qquad (X \in B(\mathcal{H}_1), \quad Y \in B(\mathcal{H}_2)) \tag{9.13}$$

is a conditional expectation; it does not depend on ρ_1.

From Takesaki's condition we obtain

$$\rho_1^{-1/2} \rho_{12}^{1/2}(X \otimes I) = (X \otimes I)\rho_1^{-1/2}\rho_{12}^{1/2}$$

for all $X \in B(\mathcal{H}_1)$. Therefore, $\rho_1^{-1/2}\rho_{12}^{1/2}$ is in the commutant and must have the form

$$\rho_1^{-1/2}\rho_{12}^{1/2} = I \otimes Y$$

for some $Y \in B(\mathcal{H}_2)$. This gives the factorization of ρ_{12}.

The example shows that the existence of a ρ_{12}-preserving conditional expectation is a very strong limitation for ρ_{12}. $\qquad\square$

The previous example can be reformulated and it has an interesting interpretation, the **no cloning theorem**.

Example 9.7. Similarly to the previous example, consider $B(\mathcal{H}_1)$ as the subsystem of $B(\mathcal{H}_1 \otimes \mathcal{H}_2)$. The state transformation $\mathcal{E} : B(\mathcal{H}_1) \rightarrow B(\mathcal{H}_1 \otimes \mathcal{H}_2)$ is a **state extension** if the reduced state of $\mathcal{E}(\rho)$ on the first subsystem is ρ itself, for every state ρ. Since the a state extension is the dual of a conditional expectation, the only possible **state extension** is

$$\mathcal{E}(\rho) = \rho \otimes \omega, \tag{9.14}$$

where ω is a fixed state of $B(\mathcal{H}_2)$.

Assume now that $\mathcal{H}_1 = \mathcal{H}_2 = \mathcal{H}$. A state transformation $\mathcal{C} : B(\mathcal{H}) \rightarrow B(\mathcal{H} \otimes \mathcal{H})$ is called **cloning** if for any input state ρ both reduced densities of the output $\mathcal{C}(\rho)$ are identical with ρ. The transformation $\mathcal{C}$ yields an output which is a pair of two subsystems, each of them in the state of the input. $\mathcal{C}$ is interpreted as copying or cloning.

Since a cloning is a particular state extension, it must have the form (9.14), which contradicts the definition of cloning (in the case when the dimension of $\mathcal{H}$ is at least two). Therefore, cloning does not exists. One can arrive at the same conclusion under the weaker assumption that $\mathcal{C}$ clones pure states only. $\qquad\square$

9.3 Commuting Squares

Let $\mathscr{A}_{123}$ be a matrix algebra with subalgebras $\mathscr{A}_{12}, \mathscr{A}_{23}, \mathscr{A}_2$ and assume that $\mathscr{A}_2 \subset \mathscr{A}_{12}, \mathscr{A}_{23}$. Then the diagram of embeddings $\mathscr{A}_2 \to \mathscr{A}_{12}, \mathscr{A}_2 \to \mathscr{A}_{23}, \mathscr{A}_{12} \to \mathscr{A}_{123}$ and $\mathscr{A}_{23} \to \mathscr{A}_{123}$ is obviously commutative (see Fig. 9.1).

Assume that a conditional expectation $E_{12}^{123} : \mathscr{A}_{123} \to \mathscr{A}_{12}$ exists. If the restriction of E_{12}^{123} to $\mathscr{A}_{23}$ is a conditional expectation to $\mathscr{A}_2$, then $(\mathscr{A}_{123}, \mathscr{A}_{12}, \mathscr{A}_{23}, \mathscr{A}_2, E_{12}^{123})$ is called a **commuting square**. The terminology comes from the commutativity of Fig. 9.2, which consists of conditional expectations and embeddings.

Theorem 9.5. *Let $\mathscr{A}_{123}$ be a matrix algebra with subalgebras $\mathscr{A}_{12}, \mathscr{A}_{23}, \mathscr{A}_2$ and assume that $\mathscr{A}_2 \subset \mathscr{A}_{12}, \mathscr{A}_{23}$. Suppose that ω_{123} is a separating state on $\mathscr{A}_{123}$ and the conditional expectations $E_{12}^{123} : \mathscr{A}_{123} \to \mathscr{A}_{12}$ and $E_2^{23} : \mathscr{A}_{23} \to \mathscr{A}_2$ preserving φ_{123} exist and $(\mathscr{A}_{123}, \mathscr{A}_{12}, \mathscr{A}_{23}, \mathscr{A}_2, E_{12}^{123})$ is a commuting square. Then the following conditions are equivalent:*

(1) $E_{12}^{123}|\mathscr{A}_{23} = E_2^{23}$.   (2) $E_{23}^{123}|\mathscr{A}_{12} = E_2^{12}$.

(3) $E_{12}^{123} E_{23}^{123} = E_{23}^{123} E_{12}^{123}$ and $\mathscr{A}_{12} \cap \mathscr{A}_{23} = \mathscr{A}_2$.

(4) $E_{12}^{123} E_{23}^{123} = E_2^{123}$.   (5) $E_{23}^{123} E_{12}^{123} = E_2^{123}$.

The idea of the proof is to consider $\mathscr{A}_{123}$ to be a Hilbert space endowed with the inner product

$$\langle A, B \rangle := \omega_{123}(A^*B) \qquad (A, B \in \mathscr{A}_{123}).$$

Then E_{12}^{123} becomes an orthogonal projection onto the subspace $\mathscr{A}_{12}$ of $\mathscr{A}_{123}$ and all conditions can be reformulated in the Hilbert space. The details will be skipped.

Example 9.8. Let $\mathscr{A}_{123} := B(\mathscr{H}_1 \otimes \mathscr{H}_2 \otimes \mathscr{H}_3)$ be the model of a quantum system consisting of three subsystems and let $\mathscr{A}_{12}, \mathscr{A}_{23}$ and $\mathscr{A}_2$ be the subsystems corresponding to the subscript, formally $\mathscr{A}_{12} := B(\mathscr{H}_1 \otimes \mathscr{H}_2) \otimes \mathbb{C}I, \mathscr{A}_{23} := \mathbb{C}I \otimes B(\mathscr{H}_1 \otimes \mathscr{H}_2)$ and $\mathscr{A}_2 := \mathbb{C}I \otimes B(\mathscr{H}_2) \otimes \mathbb{C}I$. The conditional expectations preserving the tracial state τ (or Tr) from $\mathscr{A}_{123}$ onto any subalgebra exist and unique. For example, the conditional expectation $E_{12}^{123} : \mathscr{A}_{123} \to \mathscr{A}_{12}$ is of the form

$$A \otimes B \otimes C \mapsto \tau(C) A \otimes B \otimes I.$$

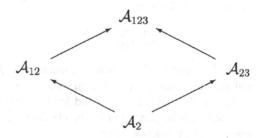

Fig. 9.1 Subalgebras of $\mathscr{A}_{123}$: $\mathscr{A}_2 \subset \mathscr{A}_{12}, \mathscr{A}_{13}$

Fig. 9.2 Commuting square, $E_{12}^{123} | \mathscr{A}_{23} = E_2^{12}$

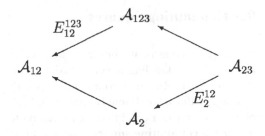

Up to a scalar, this is the partial trace over $\mathscr{H}_3$. Its restriction to $\mathscr{A}_{23}$ is

$$I \otimes B \otimes C \mapsto \tau(C) I \otimes B \otimes I$$

which is really the conditional expectation of $\mathscr{A}_{23}$ to $\mathscr{A}_2$.   □

Example 9.9. Assume that the algebra $\mathscr{A}$ is generated by the elements $\{a_i : 1 \leq i \leq n\}$, which satisfy the **canonical anticommutation relations**

$$a_i a_j + a_j a_i = 0$$
$$a_i a_j^* + a_j^* a_i = \delta_{i,j} I$$

for $1 \leq i, j \leq n$. It is known that $\mathscr{A}$ is isomorphic to a matrix algebra $M_{2^n}(\mathbb{C}) \simeq \overset{1}{M_2(\mathbb{C})} \otimes \cdots \otimes \overset{n}{M_2(\mathbb{C})}$, and the isomorphism is called **Jordan–Wigner transformation**. The relations

$$e_{11}^{(i)} := a_i a_i^*, \qquad e_{12}^{(i)} := V_{i-1} a_i,$$
$$e_{21}^{(i)} := V_{i-1} a_i^*, \qquad e_{22}^{(i)} := a_i^* a_i,$$

$$V_i := \prod_{j=1}^{i} (I - 2a_j^* a_j)$$

determine a family of mutually commuting 2×2 matrix units for $1 \leq i \leq n$. Since.

$$a_i = \prod_{j=1}^{i-1} \left(e_{11}^{(j)} - e_{22}^{(j)} \right) e_{12}^{(i)},$$

the above matrix units generate $\mathscr{A}$ and give an isomorphism between $\mathscr{A}$ and $M_2(\mathbb{C}) \otimes \cdots \otimes M_2(\mathbb{C})$:

$$e_{i_1 j_2}^{(1)} e_{i_2 j_2}^{(2)} \dots e_{i_n j_n}^{(n)} \longleftrightarrow E_{i_1 j_1} \otimes E_{i_2 j_2} \otimes \cdots \otimes E_{i_n j_n}. \tag{9.15}$$

(Here E_{ij} stand for the standard matrix units in $M_2(\mathbb{C})$.) It follows from this isomorphism that $\mathscr{A}$ has a unique tracial state.

Let $I_{12}, I_{23} \subset \{1, 2, \dots, n\}$ and let $I_2 = I_{12} \cap I_{23}$. Moreover, let $\mathscr{A}_{123} := \mathscr{A}$ and $\mathscr{A}_{12}, \mathscr{A}_{23}, \mathscr{A}_2$ be generated by the elements a_i, where i belong to the set I_{12}, I_{23}, I_2, respectively. It is known that

$$\mathscr{A}_{12} \cap \mathscr{A}_{23} = \mathscr{A}_2.$$

The trace-preserving conditional expectation $E_{12}^{123} : \mathscr{A}_{123} \to \mathscr{A}_{12}$ exists and we have a commuting square $(\mathscr{A}_{123}, \mathscr{A}_{12}, \mathscr{A}_{23}, \mathscr{A}_2, E_{12}^{123})$. □

9.4 Superadditivity

The content of the next theorem is the superadditivity of the relative entropy for a commuting square.

Theorem 9.6. *Assume that* $(\mathscr{A}_{123}, \mathscr{A}_{12}, \mathscr{A}_{23}, \mathscr{A}_2, E_{12}^{123})$ *is a commuting square,* ω_{123} *is a separating state on* $\mathscr{A}_{123}$, *and* E_{12}^{123} *leaves this state invariant. Let* ρ_{123} *be an arbitrary state on* $\mathscr{A}_{123}$ *and we denote by* $\omega_{12}, \omega_{23}, \omega_2, \rho_{12}, \rho_{23}, \rho_2$ *the restrictions of these states. Then*

$$S(\rho_{123} \| \omega_{123}) + S(\rho_2 \| \omega_2) \geq S(\rho_{12} \| \omega_{12}) + S(\rho_{23} \| \omega_{23}). \qquad (9.16)$$

Proof. The conditional expectation property of the relative entropy tells us that

$$S(\rho_{123} \| \omega) = S(\rho_{12} | \omega_{12}) + S(\rho_{123} \| \rho_{123} \circ E_{12}^{123}), \qquad (9.17)$$

$$S(\rho_2 \| \omega_2) + S(\rho_{23} \| \rho_{23} \circ E_2^{23}) = S(\rho_{23}, \omega_{23}). \qquad (9.18)$$

The monotonicity and the commuting square property give

$$S(\rho_{123} \| \rho_{123} \circ E_{12}^{123}) \geq S(\rho_{23} \| \rho_{123} \circ E_{12}^{123} | \mathscr{A}_{23}) = S(\rho_{23} \| \rho_2 \circ E_2^{23}). \qquad (9.19)$$

By adding (9.17), (9.18) and the inequality (9.19), we can conclude the statement of the theorem.

It is worthwhile to note that the necessary and sufficient condition for the equality is

$$S(\rho_{123} \| \rho_{123} \circ E_{12}^{123}) = S(\rho_{23} \| \rho_{123} \circ E_{12}^{123} | \mathscr{A}_{23}). \qquad (9.20)$$

This fact will be used later. (This condition means that the subalgebra $\mathscr{A}_{23}$ is sufficient for the states ρ_{123} and $\rho_{123} \circ E_{12}^{123}$.) □

The theorem implies the strong subadditivity of the von Neumann entropy for the tensor product structure as it is in (5.12), but a similar strong subadditivity holds for the CAR algebra as well.

9.5 Sufficiency

In order to motivate the concept of sufficiency, let us first turn to the setting of classical statistics. Suppose we observe an N-dimensional random vector $X = (x_1, x_2, \ldots, x_N)$, characterized by the density function $f(x|\theta)$, where θ is a p-dimensional vector of parameters and p is usually much smaller than N. Assume that the densities $f(x|\theta)$ are known and the parameter θ completely determines the distribution of X. Therefore, θ is to be estimated. The N-dimensional observation X carries information about the p-dimensional parameter vector θ. One may ask the following question: Can we compress x into a low-dimensional statistic without any loss of information? Does there exist some function $t = Tx$, where the dimension of t is less than N, such that t carries all the useful information about θ? If so, for the purpose of studying θ, we could discard the measurements x and retain only the low-dimensional statistic t. In this case, t is called a **sufficient statistic**. The following example is standard and simple.

Suppose a binary information source emits a sequence of -1's and $+1$'s, we have the independent variables $X_1, X_2, \ldots, X_N$ such that $\text{Prob}(X_i = 1) = \theta$. The quantum mechanical example we may have in our mind is the measurement of a Pauli observable σ_1 on N identical copies of a qubit. The empirical mean

$$T(x_1, x_2, \ldots, x_N) = \frac{1}{N} \sum_{i=1}^{N} x_i$$

can be used to estimate the parameter θ and it is a sufficient statistic. (Knowledge of the empirical mean is equivalent to the knowledge of the relative frequencies of ± 1.)

Let $\mathscr{B}$ be a matrix algebra. Assume that a family $\mathscr{S} := \{\rho_\theta : \theta \in \Theta\}$ of density matrices is given. $(\mathscr{M}, \mathscr{S})$ is called **statistical experiment**. The subalgebra $\mathscr{A} \subset \mathscr{B}$ is **sufficient** for $(\mathscr{B}, \mathscr{S})$ if there exists a coarse-graining $\alpha : \mathscr{B} \to \mathscr{A}$ such that

$$\text{Tr}\,\rho_\theta B = \text{Tr}\,\rho_\theta \alpha(B) \qquad (\theta \in \Theta, \quad B \in \mathscr{B}). \tag{9.21}$$

If we denote by $\rho_{\theta,0}$ the reduced density of ρ_θ in $\mathscr{A}$, then we can formulate sufficiency in a slightly different way. $\mathscr{A} \subset \mathscr{B}$ is **sufficient** for $(\mathscr{B}, \mathscr{S})$ if there exists a completely positive trace–preserving mapping $\mathscr{E} : \mathscr{A} \to \mathscr{B}$ such that

$$\mathscr{E}(\rho_{\theta,0}) = \rho_\theta \qquad (\theta \in \Theta). \tag{9.22}$$

Indeed, $\mathscr{E}$ satisfies (9.22) if and only if its dual α satisfies (9.21).

Before stating the main theorem characterizing sufficient subalgebras, recall the concept of the **Connes' cocycle**. If ρ and ω are density matrices, then

$$[D\rho, D\omega]_t = \rho^{it} \omega^{-it}$$

is a one-parameter family of contractions in $\mathscr{B}$ and it is called the Connes' cocycle of ρ and ω. When both densities are invertible, the Connes' cocycle consists of unitaries. The Connes' cocycle is the quantum analogue of the Radon–Nikodym derivative of measures.

The next result is called **sufficiency theorem**.

Theorem 9.7. *Let $\mathscr{A} \subset \mathscr{B}$ be matrix algebras and let $(\mathscr{B}, \{\rho_\theta : \theta \in \Theta\})$ be a statistical experiment. Assume that there are densities $\rho_n \in \mathscr{S} := \{\rho_\theta : \theta \in \Theta\}$ such that*

$$\omega := \sum_{n=1}^{\infty} \lambda_n \rho_n$$

is an invertible density for some constants $\lambda_n > 0$. Then the following conditions are equivalent.

(i) $\mathscr{A}$ is sufficient for $(\mathscr{B}, \mathscr{S})$.
(ii) $S_\alpha(\rho_\theta || \omega) = S_\alpha(\rho_{\theta,0} || \omega_0)$ for all θ and for some $0 < |\alpha| < 1$.
(iii) $[D\rho_\theta, D\omega]_t = [D\rho_{\theta,0}, D\omega_0]_t$ for every real t and for every θ.
(v) The generalized conditional expectation $E_\omega : \mathscr{B} \to \mathscr{A}$ leaves all the states ρ_θ invariant.

Lemma 9.2. Let ρ_0 and ω_0 be the reduced densities of $\rho, \omega \in \mathscr{B}$ in $\mathscr{A}$. Assume that ω is invertible. Then $S_\alpha(\rho || \omega) = S_\alpha(\rho_0 || \omega_0)$ implies

$$\rho_0^{it} \omega_0^{-it} = \rho^{it} \omega^{-it}$$

for every real number $t \in \mathbb{R}$.

Proof. The relative α-entropies can be expressed by the **relative modular operators** Δ acting on the Hilbert space $\mathscr{B}$ and Δ_0 acting on $\mathscr{A}$:

$$\Delta A = \rho A \omega^{-1} \quad (A \in \mathscr{A}) \quad \text{and} \quad \Delta_0 B = \rho_0 B \omega_0^{-1} \quad (B \in \mathscr{B}).$$

The α-entropies are

$$S_\alpha(\rho || \omega) = \frac{1}{\alpha(1 - \alpha)} (1 - \text{Tr} \, \omega^\alpha \rho^{1-\alpha}) \qquad (9.23)$$

but for the sake of simplicity we can neglect the constants:

$$S_\alpha^0(\rho || \omega) = \text{Tr} \, \omega^\alpha \rho^{1-\alpha} = \langle \omega^{1/2}, \Delta^\beta \omega^{1/2} \rangle, \qquad (9.24)$$

where $\beta = 1 - \alpha$. Assume that $0 < \alpha < 1$. (The case of negative α is treated similarly.) From the integral representation of Δ^β, we have

$$S_\alpha^0(\rho || \omega) = \frac{\sin \pi \beta}{\pi} \int_0^\infty \langle \omega^{1/2}, t^{\beta-1} \Delta(t + \Delta)^{-1} \omega^{1/2} \rangle \, dt$$

and

$$S_\alpha^0(\rho_0 || \omega_0) = \frac{\sin \pi \beta}{\pi} \int_0^\infty \langle \omega_0^{1/2}, t^{\beta-1} \Delta_0(t + \Delta_0)^{-1} \omega_0^{1/2} \rangle \, dt.$$

Due to the monotonicity of quasi-entropies, there is an inequality between the two integrands. Therefore, the equality of the entropies is equivalent to the condition

$$\langle \omega^{1/2}, \Delta(t + \Delta)^{-1} \omega^{1/2} \rangle = \langle \omega_0^{1/2}, \Delta_0(t + \Delta_0)^{-1} \omega_0^{1/2} \rangle \qquad (t > 0),$$

or

$$\langle \omega^{1/2}, (t+\Delta)^{-1}\omega^{1/2}\rangle = \langle \omega_0^{1/2}, (t+\Delta_0)^{-1}\omega_0^{1/2}\rangle \qquad (t>0). \qquad (9.25)$$

These are equalities for numbers; let us obtain equalities for operators.

The operator $V : \mathscr{A} \to \mathscr{B}$ defined as

$$VA\omega_0^{1/2} = A\omega^{1/2} \qquad (A \in \mathscr{A})$$

is an isometry and

$$V^*\Delta V \le \Delta_0.$$

The function $f_t(y) = (y+t)^{-1} - t^{-1}$ is operator monotone decreasing, so

$$(\Delta_0 + t)^{-1} - t^{-1}I \le (V^*\Delta V + t)^{-1} - t^{-1}I.$$

Moreover, f_t is operator convex and $f_t(0) = 0$, therefore

$$(V^*\Delta V + t)^{-1} - t^{-1}I \le V^*(\Delta+t)^{-1}V - t^{-1}V^*V.$$

The two equations together give

$$K := (\Delta_0 + t)^{-1} - t^{-1}I \le V^*(\Delta+t)^{-1}V - t^{-1}V^*V =: L$$

Recall that (9.25) is

$$\langle \omega_0^{1/2}, K\omega_0^{1/2}\rangle = \langle \omega_0^{1/2}, L\omega_0^{1/2}\rangle.$$

This implies that $\|(L-K)^{1/2}\omega_0^{1/2}\|^2 = 0$ and $K\omega_0^{1/2} = L\omega_0^{1/2}$, or

$$V^*(\Delta+t)^{-1}\omega^{1/2} = (\Delta_0+t)^{-1}\omega_0^{1/2} \qquad (9.26)$$

for all $t>0$. Differentiating by t we have

$$V^*(\Delta+t)^{-2}\omega^{1/2} = (\Delta_0+t)^{-2}\omega_0^{1/2}$$

and we can infer

$$\begin{aligned}
\|V^*(\Delta+t)^{-1}\omega^{1/2}\|^2 &= \langle (\Delta_0+t)^{-2}\omega_0^{1/2}, \omega_0^{1/2}\rangle \\
&= \langle V^*(\Delta+t)^{-2}\omega^{1/2}, \omega_0^{1/2}\rangle \\
&= \|(\Delta+t)^{-1}\omega^{1/2}\|^2
\end{aligned}$$

When $\|V^*\xi\| = \|\xi\|$ holds for a contraction V, it follows that $VV^*\xi = \xi$. In the light of this remark we arrive at the condition

$$VV^*(\Delta+t)^{-1}\omega^{1/2} = (\Delta+t)^{-1}\omega^{1/2}$$

and from (9.26)

$$V(\Delta_0 + t)^{-1} \omega_0^{1/2} = VV^*(\Delta + t)^{-1} \omega^{1/2}$$
$$= (\Delta + t)^{-1} \omega^{1/2}$$

The family of functions $g_t(x) = (t + x)^{-1}$ is very large and the Stone–Weierstrass approximation yields

$$V f(\Delta_0) \omega_0^{1/2} = f(\Delta) \omega^{1/2} \tag{9.27}$$

for any continuous function f. In particular for $f(x) = x^{it}$ we have

$$\rho_0^{it} \omega_0^{-it} \omega^{1/2} = \rho^{it} \omega^{-it+1/2}. \tag{9.28}$$

This condition follows from the equality of α-entropies. $\square$

Now we can prove the theorem. (i) implies (ii) due to the monotonicity of the relative α-entropies. The key point is (ii) $\Rightarrow$ (iii). This follows from Lemma 9.2.

(iii) $\Rightarrow$ (iv) is rather straightforward:

$$\mathrm{Tr}\,\rho_{\theta,0} E_\omega(B) = \mathrm{Tr}\,\rho_{\theta,0}\, \omega_0^{-1/2} F(\omega^{1/2} B \omega^{1/2}) \omega_0^{-1/2} = \mathrm{Tr}\,\rho_{\theta,0}\, \omega_0^{-1/2} \omega^{1/2} B \omega^{1/2} \omega_0^{-1/2}$$
$$= \mathrm{Tr}\,\rho_\theta\, \omega^{-1/2} \omega^{1/2} B \omega^{1/2} \omega^{-1/2} = \mathrm{Tr}\,\rho_\theta B,$$

where $\rho_{\theta,0}\, \omega_0^{-1/2} = \rho_\theta \omega^{-1/2}$ was used.

Finally, (iv) $\Rightarrow$ (i) due to the definition of sufficiency. $\square$

The density ω appearing in the theorem is said to be **dominating** for the statistical experiment $\mathscr{S}$. Given a dominated statistical experiment $\mathscr{S}$, the subalgebra generated by the operators

$$\{\rho_\theta^{it} \omega^{-it} : t \in \mathbb{R}\}$$

is the smallest sufficient subalgebra. If there are $\theta_1, \theta_2 \in \Theta$ such that ρ_{θ_1} and ρ_{θ_2} do not commute, then there exists no sufficient commutative subalgebra for $\{\rho_\theta : \theta \in \Theta\}$.

The formulation of the sufficiency theorem was made a bit complicated, since the formulation is true for an arbitrary von Neumann algebra and for a family of normal (and not necessarily faithful) states. For two invertible states in finite dimension the relative entropy is always finite and we can have a simpler formulation of the **sufficiency theorem**.

Theorem 9.8. *Let $\omega, \rho \in \mathscr{B} = B(\mathscr{H})$ be invertible density matrices on a finite-dimensional Hilbert space $\mathscr{H}$ and let $\mathscr{A} \subset \mathscr{B}$ be a subalgebra. Denote by ρ_0 and ω_0 the reduced densities in $\mathscr{A}$. Then the following conditions are equivalent.*

(i) $\mathscr{A}$ is sufficient for $\{\rho, \omega\}$.
(ii) $S_\alpha(\rho \| \omega) = S_\alpha(\rho|\mathscr{A} \| \omega|\mathscr{A})$ for some α such that $|\alpha| < 1$.
(iii) $\rho^{it} \omega^{-it} = \rho_0^{it} \omega_0^{-it}$ for every real t.
(iv) $\rho_0^{-1/2} \rho^{1/2} = \omega_0^{-1/2} \omega^{1/2}$.

(v) The generalized conditional expectations $E_\omega : \mathscr{B} \to \mathscr{A}$ and $E_\rho : \mathscr{B} \to \mathscr{A}$ coincide.

(vi) $\rho^{it} \omega^{-it} \in \mathscr{A}$ for all real t.

Proof. The equivalence (i) $\Longleftrightarrow$ (ii) $\Longleftrightarrow$ (iii) was proven in the previous theorem. (iii) $\Rightarrow$ (iv) is obvious, (iv) $\Longleftrightarrow$ (v) is Theorem 9.4. (v) $\Rightarrow$ (i) and (iii) $\Rightarrow$ (vi) are trivial.

The real problem is to prove (vi) $\Rightarrow$ (i). Let $\mathscr{A}_0$ be the algebra generated by $\{\rho^{it} \omega^{-it} : t \in \mathbb{R}\}$. Of course, $\mathscr{A}_0 \subset \mathscr{A}$ and we can write elements of $\mathscr{A}_0$ in the form

$$\bigoplus_{k=1}^{K} I_k^L \otimes A_k^R,$$

(see Sect. 11.8). We have

$$\rho^{it} \mathscr{A}_0 \rho^{-it} \subset \mathscr{A}_0$$

for every $t \in \mathbb{R}$ and Theorem 11.27 tells us that

$$\rho = \bigoplus_{k=1}^{K} A_k^L \otimes A_k^R \qquad \text{and} \qquad \omega = \bigoplus_{k=1}^{K} B_k^L \otimes B_k^R.$$

Since $\rho \omega^{-1} \in \mathscr{A}_0$, $A_k^L (B_k^L)^{-1}$ is a constant multiple of the identity, we may assume that $A_k^L = B_k^L$. The reduced densities are

$$\rho_1 = \bigoplus_{k=1}^{K} (\mathrm{Tr} A_k^L) I_k^L \otimes A_k^R \qquad \text{and} \qquad \omega_1 = \bigoplus_{k=1}^{K} (\mathrm{Tr} A_k^L) I_k^L \otimes B_k^R.$$

We can conclude that

$$\rho = \rho_1 D \qquad \text{and} \qquad \omega = \omega_1 D,$$

where $D \in \mathscr{A}_1'$. This relation implies that $\rho^{it} \omega^{-it} = \rho_1^{it} \omega_1^{-it}$ for every $t \in \mathbb{R}$. So the subalgebra $\mathscr{A}_0$ is sufficient, and the larger subalgebra $\mathscr{A}$ must be sufficient as well.

$\square$

9.6 Markov States

Let ρ_{ABC} be a density matrix acting on the finite-dimensional tensor product Hilbert space $\mathscr{H}_A \otimes \mathscr{H}_B \otimes \mathscr{H}_C$. The reduced density matrices will be denoted by ρ_{AB}, ρ_{BC} and ρ_B.

Theorem 9.9. *The following conditions are equivalent.*

(i) The equality

$$S(\rho_{ABC}) + S(\rho_B) = S(\rho_{AB}) + S(\rho_{BC})$$

holds in the strong subadditivity of the von Neumann entropy.

(ii) If τ_A denotes the tracial state of $B(\mathcal{H}_A)$, then

$$S(\rho_{ABC}\|\tau_A \otimes \rho_{BC}) = S(\rho_{AB}\|\tau_A \otimes \rho_B).$$

(iii) There exist a subalgebra $\mathcal{A}$ and a conditional expectation from $B(\mathcal{H}_A \otimes \mathcal{H}_B \otimes \mathcal{H}_C)$ onto $\mathcal{A}$ such that

$$B(\mathcal{H}_A) \otimes \mathbb{C}I_B \otimes \mathbb{C}I_C \subset \mathcal{A} \subset B(\mathcal{H}_A) \otimes B(\mathcal{H}_B) \otimes \mathbb{C}I_C$$

and E leaves the state ρ_{ABC} invariant.
(iv) The generalized conditional expectation

$$E_\rho : B(\mathcal{H}_A \otimes \mathcal{H}_B \otimes \mathcal{H}_C) \to B(\mathcal{H}_A) \otimes B(\mathcal{H}_B) \otimes \mathbb{C}I_C$$

(with respect to ρ_{ABC}) leaves the operators in $B(\mathcal{H}_A) \otimes \mathbb{C}I_B \otimes \mathbb{C}I_C$ fixed.
(v) There is a state transformation

$$\mathcal{E} : B(\mathcal{H}_B) \to B(\mathcal{H}_B \otimes \mathcal{H}_C)$$

such that $(\mathrm{id}_A \otimes \mathcal{E})(\rho_{AB}) = \rho_{ABC}$.

Proof. The equivalence (i) $\Longleftrightarrow$ (ii) is clear from the proof of the strong subadditivity in Sect. 5.4.

(iii)$\Rightarrow$(ii): To show that E leaves also the state $\tau_A \otimes \rho_{BC}$ invariant first we should establish that $E(I_A \otimes X_B \otimes X_C)$ commutes with every $X_A \otimes I_B \otimes I_C$; therefore $E(I_A \otimes X_B \otimes X_C) \in \mathbb{C}I_A \otimes B(\mathcal{H}_B) \otimes \mathbb{C}I_C$. We have

$$\begin{aligned}
(\tau_A \otimes \rho_{BC})E(X_A \otimes X_B \otimes X_C) &= (\tau_A \otimes \rho_{BC})(X_A \otimes I_B \otimes I_C)E(I_A \otimes X_B \otimes X_C) \\
&= \tau_A(X_A)\mathrm{Tr}\,\rho_{BC}E(I_A \otimes X_B \otimes X_C) \\
&= (\tau_A \otimes \rho_{BC})(X_A \otimes X_B \otimes X_C).
\end{aligned}$$

It follows that

$$S(\rho_{ABC}\|\tau_A \otimes \rho_{BC}) = S(\rho_{ABC}|\mathcal{A} \| \tau_A \otimes \rho_{BC}|\mathcal{A}).$$

Since $S(\rho_{AB}\|\tau_A \otimes \rho_B)$ is between the left-hand side and the right-hand side, it must have the same value.

(iv)$\Rightarrow$(iii): E_ρ preserves ρ_{ABC} and so do its powers. As per the Kovács–Szűcs theorem,

$$\frac{1}{n}(\mathrm{id} + E_\rho + \cdots + E_\rho^{n-1}) \to E$$

where E is a conditional expectation onto the fixed-point algebra.

(ii)$\Rightarrow$(iv) and (v): Condition (v) in Theorem 9.8 tells us that the generalized conditional expectation E_ρ has the form

$$E_\rho(X) = (\tau_A \otimes \rho_B)^{-1/2}F\left((\tau_A \otimes \rho_{BC})^{1/2}X(\tau_A \otimes \rho_{BC})^{1/2}\right)(\tau_A \otimes \rho_B)^{-1/2}$$

where F is the conditional expectation preserving the normalized trace. Let $\mathscr{A}$ be the fixed-point algebra of E_ρ. One can compute that $E_\rho(X_A \otimes I_B \otimes I_C) = X_A \otimes I_B \otimes I_C$, therefore

$$B(\mathscr{H}_A) \otimes \mathbb{C}I_B \otimes \mathbb{C}I_C \subset \mathscr{A} \subset B(\mathscr{H}_A) \otimes B(\mathscr{H}_B) \otimes \mathbb{C}I_C. \tag{9.29}$$

and (iv) is obtained. Relation (9.29) gives that $\mathscr{A}$ must be of the form $B(\mathscr{H}_A) \otimes \mathscr{A}_B \otimes \mathbb{C}I_C$ with a subalgebra $\mathscr{A}_B$ of $B(\mathscr{H}_B)$. It follows that the dual of E_ρ has the form $\mathrm{id}_A \otimes \mathscr{E}$ and we can arrive at (v).

(v) implies that $\mathscr{E}(\tau_A \otimes \rho_B) = \rho_{ABC}$ and (ii) must hold.   □

If a density matrix ρ_{ABC} acting on tensor product Hilbert space $\mathscr{H}_A \otimes \mathscr{H}_B \otimes \mathscr{H}_C$ satisfies the conditions of the previous theorem, then ρ_{ABC} will be called a **Markov state**.

If the notation

$$I(A:C|B)_\rho := S(\rho_{AB}) + S(\rho_{BC}) - S(\rho_B) - S(\rho_{ABC})$$

is used, then the Markovianity of ρ is the condition $I(A:C|B)_\rho = 0$.

Example 9.10. Assume that $\mathscr{H}_B = \mathscr{H}_L \otimes \mathscr{H}_R$. Consider densities $\rho_{AL} \in B(\mathscr{H}_A \otimes \mathscr{H}_L)$ and $\rho_{RC} \in B(\mathscr{H}_R \otimes \mathscr{H}_C)$ and let $\rho_{ABC} = \rho_{AL} \otimes \rho_{RC}$. Then ρ_{ABC} is a Markov state, and we shall call it "a product type." It is easy to check the strong additivity of the von Neumann entropy:

$$S(\rho_{ABC}) = S(\rho_{AL}) + S(\rho_{RC}), \quad S(\rho_B) = S(\rho_L) + S(\rho_R),$$
$$S(\rho_{AB}) = S(\rho_{AL}) + S(\rho_R), \quad S(\rho_{BC}) = S(\rho_L) + S(\rho_{RC}).$$

Let P and Q be orthogonal projections. Assume that ρ_{ABC} and ω_{ABC} are Markov states with support in P and Q, respectively. Computation of the von Neumann entropies yields that any convex combination $\lambda \rho_{ABC} + (1-\lambda)\omega_{ABC}$ is a Markov state as well.

It is the content of the next theorem that every Markov state is the convex combination of orthogonal product type states.   □

Theorem 9.10. *Assume that ρ_{ABC} is a Markov state on the finite dimensional tensor product Hilbert space $\mathscr{H}_A \otimes \mathscr{H}_B \otimes \mathscr{H}_C$. Then $\mathscr{H}_B$ has an orthogonal decomposition*

$$\mathscr{H}_B = \bigoplus_k \mathscr{H}_k^L \otimes \mathscr{H}_k^R$$

and for every k there are density matrices $\rho_{AL}^k \in B(\mathscr{H}_A \otimes \mathscr{H}_k^L)$ and $\rho_{RC}^k \in B(\mathscr{H}_k^R \otimes \mathscr{H}_C)$ such that ρ_{ABC} is a convex combination

$$\rho_{ABC} = \sum_k p_k \rho_{AL}^k \otimes \rho_{RC}^k.$$

Proof. Let $\omega := \tau_A \otimes \rho_{BC}$ and $\rho := \rho_{ABC}$. We know from Theorem 9.9 that Markovianity implies that the generalized conditional expectation

$$E_\omega : B(\mathcal{H}_A) \otimes B(\mathcal{H}_B) \otimes B(\mathcal{H}_C) \to B(\mathcal{H}_A) \otimes B(\mathcal{H}_B) \otimes \mathbb{C}I_C$$

with respect to the state ω is the same as E_ρ and the fixed-point algebra $\mathscr{A}$ has the property

$$B(\mathcal{H}_A) \otimes \mathbb{C}I_B \otimes \mathbb{C}I_C \subset \mathscr{A} \subset B(\mathcal{H}_A) \otimes B(\mathcal{H}_B) \otimes \mathbb{C}I_C.$$

It follows that $\mathscr{A}$ must be of the form $B(\mathcal{H}_A) \otimes \mathscr{A}_B \otimes \mathbb{C}I_C$ with a subalgebra $\mathscr{A}_B$ of $B(\mathcal{H}_B)$. Elements of $\mathscr{A}_B$ have the form

$$\bigoplus_{k=1}^{K} A_k^L \otimes I_k^R,$$

where $A_k^L \in B(\mathcal{H}_k^L)$, I_k^R is the identity on $\mathcal{H}_k^R$ and

$$\mathcal{H}_B = \bigoplus_{k=1}^{K} \mathcal{H}_k^L \otimes \mathcal{H}_k^R.$$

In this way $\mathscr{A}$ is isomorphic to

$$\bigoplus_{k=1}^{K} B(\mathcal{H}_A \otimes \mathcal{H}_k^L) \otimes \mathbb{C}I_k^R \otimes \mathbb{C}I_C.$$

Since $\rho^{it} \mathscr{A} \rho^{-it} \subset \mathscr{A}$ holds, Theorem 11.27 can be applied and gives the stated decomposition. □

9.7 Notes

The conditional expectation in the matrix algebra (or von Neumann algebra) setting was introduced by Umegaki in [115] and Example 9.5 is due to him. Takesaki's theorem was developed in 1972. In the infinite-dimensional von Neumann algebra case, the modular operators are unbounded and condition (9.9) is replaced by the equation

$$\Delta_\mathscr{A}^{it} A \Delta_\mathscr{A}^{-it} = \Delta_\mathscr{B}^{it} A \Delta_\mathscr{B}^{-it}$$

for real t's. (Note that $\Delta_\mathscr{A}^{it}$ and $\Delta_\mathscr{B}^{it}$ are unitaries and the operators $\Delta_\mathscr{A}^z$ and $\Delta_\mathscr{B}^z$ are not everywhere defined for a complex $z \in \mathbb{C}$.) The generalized conditional expectation was introduced by Accardi and Cecchini, [90] is a suggested survey.

Example 9.4 is a particular case of the **Kovács–Szűcs theorem** which concerns a semigroup of coarse-grainings in von Neumann algebras. That was the first ergodic theorem in the von Neumann algebra setting.

In general von Neumann algebras, the **Connes' cocycle** is defined in terms of relative modular operators:

$$[D\psi, D\omega]_t = \Delta(\psi/\omega)^{it}\Delta(\omega/\omega)^{-it}$$

is a one-parameter family of contractions in the von Neumann algebra, although the relative modular operators $\Delta(\psi/\omega)$ are typically unbounded.

The **conditional expectation property** together with invariance, direct sum property, nilpotence and measurability can be used to axiomatize the relative entropy (see Chap. 2 of [83]). The conditional expectation property in full generality is Theorem 5.15 in the same monograph.

Sufficiency in the quantum setting was initiated by Petz in [89] for subalgebras and in [91] for coarse-grainings. Recent paper on the subject is [63], and [64] is a survey with examples.

The concept of Markov state goes back to Accardi and Frigerio [1]. The original definition was formulated in terms of generalized or quasi-conditional expectation (see also Lemma 11.3 in [83]).

Theorem 9.10 is due to Hayden et al. [47], the presented proof follows [76]. The theorem holds in infinite-dimensional Hilbert space if all the von Neumann entropies $S(\rho_{AB}), S(\rho_B), S(\rho_{BC})$ are finite (see [63]). The Markov property in the quantum setting is an interesting subject to study (see [61], for example).

9.8 Exercises

1. Let $\mathscr{A} \subset \mathscr{B}$ be matrix algebras, $\rho \in \mathscr{B}$ be a density matrix with reduced density $\rho_0 \in \mathscr{A}$. Endow $\mathscr{A}$ with the inner product $\langle A_1, A_2 \rangle := \mathrm{Tr}\, A_1^* \rho_0^{1/2} A_2 \rho_0^{1/2}$ and similarly let $\langle B_1, B_2 \rangle := \mathrm{Tr}\, B_1^* \rho^{1/2} B_2 \rho^{1/2}$ be an inner product on $\mathscr{B}$. Show that the adjoint of the embedding $\mathscr{A} \to \mathscr{B}$ is the **generalized conditional expectation** E_ρ defined in (9.10).

2. Let $\mathscr{B}$ be a matrix algebra and $\rho \in \mathscr{B}$ be an invertible density matrix. Assume that $\alpha : \mathscr{B} \to \mathscr{B}$ is a coarse-graining leaving the state ρ invariant and let $\mathscr{A}$ be the set of fixed points of α. Show that for every $B \in \mathscr{B}$ the set

$$\overline{\mathrm{conv}}\{\alpha^n(B) : n \in \mathbb{Z}^+\} \cap \mathscr{A}$$

is a singleton.

Chapter 10
State Estimation

The goal of state estimation is to determine the density operator ρ of a quantum system by measurements on n copies of the quantum system which are all prepared in the state ρ. Since the result of a measurement in quantum mechanics is random, several measurements should be made to get information. A measurement changes the state of the system drastically; therefore an identically prepared other system is used for the next measurement. The number n corresponds to the sample size in classical mathematical statistics. An estimation scheme consists of a measurement and an estimate for every n, and the estimation error is expected to tend to 0 when n tends to infinity. It is also possible that the aim of the estimation is not the density matrix itself but certain function of the density matrix.

10.1 Estimation Schemas

Let Θ be a set of density matrices acting on a Hilbert space $\mathcal{H}$. For each $n \in \mathbb{N}$ a positive-operator-valued measure $F_n : \mathcal{B}(X_n) \to B(\mathcal{H}^{\otimes n})$ is given on the Borel sets of $\mathcal{X}_n$. This means that

(a) for $H \subset \mathcal{X}_n$ a positive (self-adjoint) operator $F_n(H)$ acting on $\mathcal{H}^{\otimes n}$ is given,
(b) if $H_1, H_2, \cdots \subset \mathcal{X}_n$ are pairwise disjoint, then

$$F_n(\cup_k H_k) = \sum_k F_n(H_k),$$

(c) $F_n(\mathcal{X}_n) = I$.

Such an F_n is regarded as a collective measurement with values in $\mathcal{X}_n$. Sometimes $\mathcal{X}_n$ is a finite set and we have a partition of the identity. When all operators $F_n(H)$ are projections, we have another important special case which is called **simple** or **von Neumann measurement**.

The probability that for a given state $\rho \in \Theta$ the measurement value is in $H \subset \mathcal{X}_n$ is given by

D. Petz, *State Estimation*. In: D. Petz, Quantum Information Theory and Quantum Statistics, Theoretical and Mathematical Physics, pp. 143–164 (2008)
DOI 10.1007/978-3-540-74636-2_10

$$\mu_{n,\rho}(H) = \text{Tr}\,\rho^{\otimes n} F_n(H) \tag{10.1}$$

which is the so-called **Bohr's rule**.

Beyond the measurement an **estimator** $\Phi_n : \mathscr{X}_n \to \Theta$ is given. When the value of the measurement is $x \in \mathscr{X}_n$, we can infer that the state of the system is $\Phi_n(x)$. The sequence (F_n, Φ_n) of pairs is called **estimation scheme**. The estimation scheme is called **unbiased** if the expectation value of Φ_n under $\mu_{n,\rho}$ is ρ for every $\rho \in \Theta$. The estimation scheme (F_n, Φ_n) is called **consistent** if for every $\rho \in \Theta$ the distribution of the random variable Φ_n converges weakly to the point-measure concentrated on ρ as $n \to \infty$.

Example 10.1. Fix a vector w on the Bloch sphere. Then

$$P_{\pm} := \tfrac{1}{2}(\sigma_0 \pm w \cdot \sigma)$$

are orthogonal projections and $P_+ + P_- = I$ follows. They form a von Neumann measurement with values ± 1: $\mathscr{X} = \{-1, 1\}$, $F(-1) = P_-$ and $F(1) = P_+$.

For the density matrix

$$\rho_\theta = \tfrac{1}{2}(\sigma_0 + \theta(w \cdot \sigma))$$

$(-1 \leq \theta \leq 1)$, Bohr's rule gives the probabilities

$$p = \text{Tr}\,\rho_\theta P_+ = \frac{1 + \theta}{2} \quad \text{and} \quad 1 - p = \text{Tr}\,\rho_\theta P_- = \frac{1 - \theta}{2},$$

hence the probability measure on $\{-1, 1\}$ is $(1 - p, p)$. (This measure depends on the parameter θ.)

Let the estimator $\Phi : \{-1, 1\} \to \Theta = \{\rho_\theta : -1 \leq \theta \leq 1\}$ be given as

$$\Phi(\pm 1) = \tfrac{1}{2}(\sigma_0 \pm (w \cdot \sigma)).$$

So Φ always gives a pure state. The expectation value of the estimate is

$$p\Phi(1) + (1 - p)\Phi(-1) = \tfrac{1}{2}(\sigma_0 + \theta(x \cdot \sigma)) = \rho_\theta.$$

Therefore, this estimation is unbiased.

Note that $P_{\pm}$ are the spectral projections of the matrix $w \cdot \sigma$, which is the observable corresponding to the spin in the direction w. What we discussed is not a full estimation scheme, we had a single measurement only. $\square$

In the previous example the set Θ was parameterized with a single real parameter. Very often $\Theta = \{\rho_\theta : \theta = (\theta_1, \theta_2, \ldots, \theta_N) \in G\}$, where $G \subset \mathbb{R}^N$. When the state ρ_θ is identified with the N-tuple $(\theta_1, \theta_2, \ldots, \theta_N)$, the task of the state estimation is to determine the correct values of $\theta_1, \theta_2, \ldots, \theta_N$.

Example 10.2. Assume that Θ is a convex set of density matrices acting on a Hilbert space $\mathscr{H}$, a positive-operator-valued measure $F : \mathscr{B}(\mathscr{X}) \to B(\mathscr{H})$ and an estimator

$\Phi : \mathcal{X} \to \Theta$ are given in the single measurement case. There is a very natural way to construct a full state estimation scheme. Take

$$\mathcal{X}_n := \mathcal{X} \times \mathcal{X} \times \ldots \times \mathcal{X},$$
$$F_n := F \otimes F \otimes \ldots \otimes F,$$
$$\Phi_n(x_1, x_2, \ldots, x_n) := \frac{1}{n}(\Phi(x_1) + \Phi(x_2) + \cdots + \Phi(x_n)).$$

Then

$$\mu_{n,\rho} = \mu_\rho \otimes \mu_\rho \otimes \ldots \otimes \mu_\rho, \tag{10.2}$$

where μ_ρ is the distribution of the measurement F in the state ρ_θ. (In the background, we have n independent measurements.)

If the expectation value of Φ is the true value of θ, then this estimation scheme is unbiased. Moreover, the law of large numbers tells us that Φ_n converges to the mean of Φ. Therefore the estimation scheme is consistent. □

Example 10.3. Assume that the measures $\mu_{n,\rho}$ are absolute continuous with respect to a dominating measure and $d\mu_{n,\rho}(x) = f_{n,\rho}(x)\,dx$ for some density functions $f_{n,\rho}(x)$. Assume that the actual value of the measurement is $x \in \mathcal{X}_n$. Then the **maximum likelihood estimate** $\Phi_n^{ML}(x) \in \Theta$ is the maximizer of the function

$$\rho \mapsto f_{n,\rho}(x),$$

that is, $\Phi_n^{ML}(x) = \mathrm{argmax}_\rho f_{n,\rho}(x)$. Since the logarithm is a monotone function, we can equivalently take

$$\Phi_n^{ML}(x) = \mathrm{argmax}_\rho \log f_{n,\rho}(x). \tag{10.3}$$

□

Example 10.4. Now our aim is to extend Example 10.1 in such a way that it should cover the reconstruction of a state of a qubit. Θ will be the full state space represented by the Bloch ball.

For $i = 1, 2, 3$, let

$$P_i := \tfrac{1}{2}(\sigma_0 + \sigma_i) \quad \text{and} \quad Q_i := \tfrac{1}{2}(\sigma_0 - \sigma_i).$$

These are orthogonal projections and $P_i + Q_i = I$. (The spectral decomposition of σ_i is $P_i - Q_i$.) Similarly to Example 10.1, we have three projection-valued measures on $\{-1, 1\}$ and we can form their product:

$$F := (\delta_{-1}Q_1 + \delta_1 P_1) \otimes (\delta_{-1}Q_2 + \delta_1 P_2) \otimes (\delta_{-1}Q_3 + \delta_1 P_3)$$

is a positive-operator-valued measure from $\mathcal{X} := \{-1, 1\}^3 \to B(\mathbb{C}^8)$. ($\delta_x$ stands for the Dirac measure concentrated on x.) F is a von Neumann measurement on three qubits.

For the density matrix

$$\rho_\theta = \tfrac{1}{2}(\sigma_0 + \theta \cdot \sigma)$$

$(\theta \in \mathbb{R}^3, \|\theta\|_2 \le 1)$, the Bohr's rule gives the probabilities

$$p_i = \mathrm{Tr}\,\rho_\theta P_i = \frac{1+\theta_i}{2} \quad \text{and} \quad 1-p_i = \mathrm{Tr}\,\rho_\theta Q_i = \frac{1-\theta_i}{2},$$

and the a posteriori probability measure is

$$\mu_\theta = \otimes_{i=1}^3 ((1-p_i)\delta_{-1} + p_i\delta_1).$$

Assume that the result of the measurement is $(\varepsilon_1, \varepsilon_2, \varepsilon_3)$, where $\varepsilon_i = \pm 1$. In order to make a maximum likelihood estimate, we have to maximize

$$\frac{1}{8}\prod_{i=1}^3 (1+\varepsilon_i\theta_i)$$

on the Bloch ball. Due to symmetry, this can be done without computation. The maximum likelihood estimate

$$\Phi(\varepsilon_1, \varepsilon_2, \varepsilon_3) = \frac{1}{\sqrt{3}}(\varepsilon_1, \varepsilon_2, \varepsilon_3)$$

always gives a pure state. The Bloch ball cannot be the convex hull of finitely many points; this implies that the estimator is not unbiased.   □

In the next example the state of a qubit is estimated by means of a measurement of four outcomes.

Example 10.5. Consider the following Bloch vectors

$$a_1 = \frac{1}{\sqrt{3}}(1,1,1), \qquad a_2 = \frac{1}{\sqrt{3}}(1,-1,-1),$$

$$a_3 = \frac{1}{\sqrt{3}}(-1,1,-1), \quad a_4 = \frac{1}{\sqrt{3}}(-1,-1,1).$$

and form the positive operators

$$F_i = \frac{1}{4}(\sigma_0 + a_i \cdot \sigma) \qquad (1 \le i \le 4).$$

They determine a measurement, $\sum_{i=1}^4 F_i = I$. The probability of the outcome i is

$$p_i = \mathrm{Tr}\,F_i\rho_\theta = \frac{1}{4}(1 + a_i \cdot \theta).$$

The maximum likelihood estimate is

$$\Phi(i) = \frac{1}{2}(\sigma_0 + a_i \cdot \sigma).$$

$\square$

Now let us continue Example 10.4 and extend it to a full estimation scheme.

Example 10.6. The result of the measurement in Example 10.4 is a triplet, $(\varepsilon_1, \varepsilon_2, \varepsilon_3)$, where $\varepsilon_i = \pm 1$. Remember that ε_i is the measured value of the spin in direction i. Assume that the measurement is performed n times on identical but different systems. Then the measurement data is a sequence

$$x := (\varepsilon_1^{(1)}, \varepsilon_2^{(1)}, \varepsilon_3^{(1)}, \varepsilon_1^{(2)}, \varepsilon_2^{(2)}, \varepsilon_3^{(2)}, \dots, \varepsilon_1^{(n)}, \varepsilon_2^{(n)}, \varepsilon_3^{(n)}).$$

Assume that the true state is ρ_θ. Then

$$\mu_n(x) = \prod_{i=1}^{3} p_i^{n_i(x)} (1 - p_i)^{n - n_i(x)}, \quad \text{where} \quad n_i(x) = \#\{1 \le j \le n : \varepsilon_i^{(j)} = 1\}.$$

In order to estimate, we do not actually need the full sequence x, we may benefit from a sufficient statistics. What we need is the relative frequencies:

$$v_i^{(n)}(x) := \frac{n_i(x)}{n}$$

According to the law of large numbers $v_i^{(n)}(x) \to p_i = (1 + \theta_i)/2$. Therefore,

$$\tilde{\Phi}_n(x) := \left((2v_1^{(n)}(x) - 1),\ (2v_2^{(n)}(x) - 1),\ (2v_3^{(n)}(x) - 1) \right) \tag{10.4}$$

seems to be a good estimate of θ. Let us compute the expectation value

$$\sum_x \tilde{\Phi}_n(x) \prod_{i=1}^{3} p_i^{n_i(x)} (1 - p_i)^{n - n_i(x)}.$$

The ith component is

$$\sum_x p_i^{n_i(x)} (1 - p_i)^{n - n_i(x)} (2v_i^{(n)}(x) - 1) = \sum_{k=0}^{n} \binom{n}{j} p_i^k (1 - p_i)^{n-k} \left(\frac{2j}{n} - 1 \right) = 2p_i - 1 = \theta_i.$$

The expectation value of $\tilde{\Phi}_n$ is the true state. We can arrive at the same conclusion in a different way. Observe that

$$\tilde{\Phi}_n(x) = \frac{1}{n} \sum_{i=1}^{n} \tilde{\Phi}_1(\varepsilon_1^{(i)}, \varepsilon_2^{(i)}, \varepsilon_3^{(i)}).$$

Since $\tilde{\Phi}_1$ is unbiased, $\tilde{\Phi}_n$ is unbiased as well (cf. Example 10.2).

It may happen that the value of $\tilde{\Phi}_n$ is outside of the Bloch ball. (For $n = 1$ this is always the case.) Therefore, we can modify $\tilde{\Phi}_n$ as follows:

$$\Phi_n(x) = \begin{cases} \tilde{\Phi}_n(x) & if \|\tilde{\Phi}_n(x)\| \le 1, \\[2ex] \dfrac{\tilde{\Phi}_n(x)}{\|\tilde{\Phi}_n(x)\|} & \text{otherwise.} \end{cases}$$

This modification is natural on the one hand, and it is justified by the method of **least squares** on the other hand. Given the observation data $(v_1^{(n)}(x), v_2^{(n)}(x), v_3^{(n)}(x))$, minimization of the measure of fit,

$$\sum_{i=1}^{3} \left(v_i^{(n)}(x) - \frac{1+\theta_i}{2} \right)^2,$$

yields exactly $\Phi_n(x)$.

To show that Φ_n is an asymptotically unbiased estimator we should study the difference

$$\sum_x \tilde{\Phi}_n(x) p(x) - \sum_x \Phi_n(x) p(x) = \sum_{x \in D} (\tilde{\Phi}_n(x) - \Phi_n(x)) p(x), \qquad (10.5)$$

where the latest summation is over all $x \in \mathcal{X}_n$ such that

$$\sum_i \left(2 v_i^{(n)}(x) - 1 \right)^2 = \|\tilde{\Phi}_n(x)\|^2 > 1. \qquad (10.6)$$

If the true state is mixed, then the probability of (10.6) tends to 0 as $n \to \infty$ according to the law of large numbers. We can conclude that Φ_n is asymptotically unbiased. If the true state is pure, we need a bit longer argument. Let us divide D into two subsets

$$D_1 := \{x \in D : \|\tilde{\Phi}_n(x)\| > 1 + \varepsilon\} \quad \text{and} \quad D_2 := \{x \in D : 1 < \|\tilde{\Phi}_n(x)\| \le 1 + \varepsilon\}.$$

Then

$$\sum_{x \in D} \|\tilde{\Phi}_n(x) - \Phi_n(x)\| p(x) \le \sum_{x \in D_1} \|\tilde{\Phi}_n(x) - \Phi_n(x)\| p(x) + \sum_{x \in D_2} \|\tilde{\Phi}_n(x) - \Phi_n(x)\| p(x).$$

The first term is majorized by $2\mathrm{Prob}(D_1)$ and the second one by ε. Since the first tends to 0 and the latter is arbitrarily small, we can conclude that (10.5) tends to 0.

Let G be an open set such that $\theta \in G$. According to the law of large numbers

$$\mathrm{Prob}(\tilde{\Phi}_n(x) \notin G) \to 0,$$

however, according to the **large deviation theorem** the convergence is exponentially fast:

$$\mathrm{Prob}(\tilde{\Phi}_n(x) \notin G) \le C \exp(-n E_G),$$

where the exponent is the infimum of the so-called **rate function**:

$$E_G := \inf\left\{ \sum_{i=1}^{3} S\Big([(1+t_i)/2,(1-t_i)/2)]\Big\|[(1+\theta_i)/2,(1-\theta_i)/2]\Big) : t \notin G \right\}.$$

$E_G > 0$, since the rate function is the sum of relative entropies and it equals 0 if and only if $t_i = \theta_i$, which is excluded by the condition $t \notin G$. (I do not want to justify here the concrete form of the rate function.)

The exponential convergence tells us that $\check{\Phi}_n$ violates the constraint $\|\check{\Phi}_n\| \le 1$ with very small probability if n is large and $\|\theta\| < 1$. $\qquad\square$

Example 10.7. If we have a k-level quantum system, then its density matrix has $k^2 - 1$ real parameters. There are many ways to construct an estimator such that its values are self-adjoint $k \times k$ matrices of trace 1. We can call this $\check{\Phi}$ **unconstrained estimator**, since its values are not always positive semidefinite. We can modify $\check{\Phi}$ to get really an estimator in the following way:

$$\Phi := \mathrm{argmin}_\omega \mathrm{Tr}(\check{\Phi} - \omega)^2 = \mathrm{argmin}_\omega \sum_{i,j} (\check{\Phi})_{ij} - \omega_{ij})^2, \qquad (10.7)$$

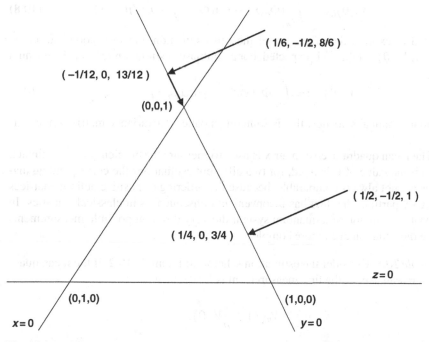

Fig. 10.1 Modification of the unconstrained estimate for 3×3 matrices is shown on the plane $x+y+z = 1$ of $\mathbb{R}^3$. The triangle $\{(x,y,z) : x,y,z \ge 0\}$ corresponds to the diagonal density matrices. Starting from the unconstrained estimate $\mathrm{Diag}(1/2,-1/2,1)$, the constrained $\mathrm{Diag}(1/4,0,3/4)$ is reached in one step. Starting from $\mathrm{Diag}(1/6,-1/2,8/6)$, two steps are needed

where ω runs over the density matrices. The $k \times k$ density matrices form a closed convex set $\mathscr{D}_k$; therefore the minimizer is unique.

We can change the basis such that $\tilde{\Phi}$ becomes diagonal, since the Hilbert–Schmidt distance is invariant under this transformation. So let $U\tilde{\Phi}U^* = \mathrm{Diag}\,(x_1, x_2, \ldots, x_k)$ for a unitary U. Then we can compute the minimizer in (10.7) by the following algorithm. Assume that the sum of the negative entries is $C < 0$. We can replace the ℓ negative entries by 0 and add $C/(k - \ell)$ to the other entries. In this way the trace remains 1. If there is no negative entry, then we are ready, otherwise we have to repeat the above procedure. After finitely many steps there is no negative entry (see Fig. 10.1).   □

10.2 Cramér–Rao Inequalities

The Cramér–Rao inequality belongs to the basics of estimation theory in mathematical statistics.

Assume that we have to estimate the state ρ_θ, where $\theta = (\theta_1, \theta_2, \ldots, \theta_N)$ lies in a subset of $\mathbb{R}^N$. In mathematical statistics the $N \times N$ **mean quadratic error matrix**

$$V_n(\theta)_{i,j} := \int_{\mathscr{X}_n} (\Phi_n(x)_i - \theta_i)(\Phi_n(x)_j - \theta_j)\,d\mu_{n,\theta}(x) \tag{10.8}$$

is used to express the efficiency of the nth estimation and in a good estimation scheme $V_n(\theta) = O(n^{-1})$ is expected. For an unbiased estimation scheme the formula simplifies:

$$V_n(\theta)_{i,j} := \int_{\mathscr{X}_n} \Phi_n(x)_i \Phi_n(x)_j\,d\mu_{n,\theta}(x) - \theta_i\theta_j,. \tag{10.9}$$

(In mathematical statistics, this is sometimes called "covariance matrix of the estimate.")

The mean quadratic error matrix is used to measure the efficiency of an estimate. Even if the value of θ is fixed, for two different estimations the corresponding matrices are not always comparable, because the ordering of positive definite matrices is highly partial. This fact has inconvenient consequences in classical statistics. In the state estimation of a quantum system the very different possible measurements make the situation even more complex.

Example 10.8. Consider the estimation scheme of Example 10.2. If the mean quadratic error matrix of the first measurement is $V(\theta)$, then

$$V_n(\theta) = \frac{1}{n}V(\theta).$$

This follows from the well-known fact that for a sequence of independent (vector-valued) random variables $\xi_1, \xi_2, \ldots, \xi_N$ the covariance matrix of $\eta := (\xi_1 + \xi_2 + \cdots + \xi_N)/N$ is the mean of the individual covariance matrices:

$$\mathrm{Cov}\,(\eta) = \frac{1}{n}\sum_{i=1}^{N}\mathrm{Cov}\,(\xi_i)$$

We have the convergence $V_n \to 0$ as $n \to \infty$. $\qquad\qquad\qquad\qquad\square$

Assume that $d\mu_{n,\theta}(x) = f_{n,\theta}(x)\,dx$ and fix θ. $f_{n,\theta}$ is called **likelihood function**. Let

$$\partial_j = \frac{\partial}{\partial\theta_j}.$$

Differentiating the relation

$$\int_{\mathscr{X}_n} f_{n,\theta}(x)\,dx = 1,$$

we have

$$\int_{\mathscr{X}_n} \partial_j f_{n,\theta}(x)\,dx = 0.$$

If the estimation scheme is unbiased, then

$$\int_{\mathscr{X}_n} \Phi_n(x)_i \partial_j f_{n,\theta}(x)\,dx = \delta_{i,j}.$$

As a combination, we can conclude

$$\int_{\mathscr{X}_n} (\Phi_n(x)_i - \theta_i)\partial_j f_{n,\theta}(x)\,dx = \delta_{i,j}$$

for every $1 \le i,j \le N$. This condition may be written in the slightly different form

$$\int_{\mathscr{X}_n} \left((\Phi_n(x)_i - \theta_i)\sqrt{f_{n,\theta}(x)}\right)\frac{\partial_j f_{n,\theta}(x)}{\sqrt{f_{n,\theta}(x)}}\,dx = \delta_{i,j}.$$

Now the first factor of the integrand depends on i while the second one on j. We need the following lemma.

Lemma 10.1. *Assume that u_i, v_i are vectors in a Hilbert space such that*

$$\langle u_i, v_j \rangle = \delta_{i,j} \qquad (i,j = 1,2,\ldots,N).$$

Then the inequality

$$A \ge B^{-1}$$

holds for the $N \times N$ matrices

$$A_{i,j} = \langle u_i, u_j \rangle \quad and \quad B_{i,j} = \langle v_i, v_j \rangle \qquad (1 \le i,j \le N).$$

The lemma applies to the vectors

$$u_i = (\Phi_n(x)_i - \theta_i)\sqrt{f_{n,\theta}(x)} \quad \text{and} \quad v_j = \frac{\partial_j f_{n,\theta}(x)}{\sqrt{f_{n,\theta}(x)}}$$

and the matrix A will be exactly the mean square error matrix $V_n(\theta)$, while in place of B we have

$$\mathbf{I}_n(\theta)_{i,j} = \int_{\mathscr{X}_n} \frac{\partial_i(f_{n,\theta}(x))\partial_j(f_{n,\theta}(x))}{f_{n,\theta}^2(x)}\, d\mu_{n,\theta}(x).$$

Therefore, the lemma tells us the following.

Theorem 10.1. *For an unbiased estimation scheme the matrix inequality*

$$V_n(\theta) \geq \mathbf{I}_n(\theta)^{-1}$$

holds (if the likelihood functions $f_{n,\theta}$ satisfy certain regularity conditions).

This is the classical **Cramér–Rao inequality**. The right-hand side is called **Fisher information matrix**. The essential content of the inequality is that the lower bound is independent of the estimate Φ_n but depends on the the classical likelihood function. The inequality is called "classical" because on both sides classical statistical quantities appear.

Example 10.9. Let F be a measurement with values in the finite set $\mathscr{X}$ and assume that $\rho_\theta = \rho + \sum_{i=1}^n \theta_i B_i$. Let us compute the Fisher information matrix at $\theta = 0$.
Since

$$\partial_i \mathrm{Tr}\, \rho_\theta F(x) = \mathrm{Tr}\, B_i F(x)$$

for $1 \leq i \leq n$ and $x \in \mathscr{X}$, we have

$$\mathbf{I}_{ij}(0) = \sum_{x \in \mathscr{X}} \frac{\mathrm{Tr}\, B_i F(x)\mathrm{Tr}\, B_j F(x)}{\mathrm{Tr}\, \rho F(x)}$$

□

When the estimation scheme of Example 10.2 is considered, we have $\mathbf{I}_n(\theta) = n\mathbf{I}(\theta)$ and the inequality becomes

$$V_n(\theta) \geq \frac{1}{n}\mathbf{I}(\theta)^{-1}.$$

The essential point in the quantum Cramér–Rao inequality compared with Theorem 10.1 is that the lower bound is a quantity determined by the family Θ. Theorem 10.1 allows to compare different estimates for a given measurement but two different measurements are not comparable.

As a starting point I give a very general form of the quantum Cramér–Rao inequality in the simple setting of a single parameter. For $\theta \in (-\varepsilon, \varepsilon) \subset \mathbb{R}$ a statistical operator ρ_θ is given and the aim is to estimate the value of the parameter θ close to 0. Formally ρ_θ is an $m \times m$ positive semidefinite matrix of trace 1 which describes a mixed state of a quantum mechanical system and it is assumed that ρ_θ is smooth (in θ). Assume that an estimation is performed by the measurement of a self-adjoint matrix A playing the role of an observable. (In this case the positive-operator-valued measure on $\mathbb{R}$ is the spectral measure of A.) A is unbiased estimator when $\mathrm{Tr}\,\rho_\theta A = \theta$. Assume that the true value of θ is close to 0. A is called **locally unbiased estimator** (at $\theta = 0$) if

$$\frac{\partial}{\partial \theta}\mathrm{Tr}\,\rho_\theta A\bigg|_{\theta=0} = 1. \tag{10.10}$$

Of course, this condition holds if A is an unbiased estimator for θ. To require $\mathrm{Tr}\,\rho_\theta A = \theta$ for all values of the parameter might be a serious restriction on the observable A and therefore it is preferred to use the weaker condition (10.10).

Example 10.10. Let

$$\rho_\theta := \frac{\exp(H + \theta B)}{\mathrm{Tr}\,\exp(H + \theta B)}$$

and assume that $\rho_0 = e^H$ is a density matrix and $\mathrm{Tr}\,e^H B = 0$. The Frechet derivative of ρ_θ (at $\theta = 0$) is $\int_0^1 e^{tH} B e^{(1-t)H}\,dt$. Hence the self-adjoint operator A is locally unbiased if

$$\int_0^1 \mathrm{Tr}\,\rho_0^t B \rho_0^{1-t} A\,dt = 1.$$

(Note that ρ_θ is a quantum analogue of the **exponential family**; in terms of physics ρ_θ is a **Gibbsian family** of states.) □

Let $\varphi_\rho[B,C] = \mathrm{Tr}\,\mathbb{J}_\rho(B)C$ be an inner product on the linear space of self-adjoint matrices. $\varphi_\rho[\cdot,\cdot]$ and the corresponding super-operator $\mathbb{J}_\rho$ depend on the density matrix ρ; the notation reflects this fact. When ρ_θ is smooth in θ, as was already assumed, then

$$\frac{\partial}{\partial \theta}\mathrm{Tr}\,\rho_\theta B\bigg|_{\theta=0} = \varphi_{\rho_0}[B,L] \tag{10.11}$$

with some $L = L^*$. From (10.10) and (10.11), we have $\varphi_0[A,L] = 1$ and the Schwarz inequality yields the following.

Theorem 10.2.

$$\varphi_{\rho_0}[A,A] \geq \frac{1}{\varphi_{\rho_0}[L,L]}. \tag{10.12}$$

This is the **quantum Cramér–Rao inequality** for a locally unbiased estimator. It is instructive to compare Theorem 10.2 with the classical Cramér–Rao inequality. If $A = \sum_i \lambda_i E_i$ is the spectral decomposition, then the corresponding von Neumann measurement is $F = \sum_i \delta_{\lambda_i} E_i$. Take the estimate $\Phi(\lambda_i) = \lambda_i$. Then the mean quadratic error is $\sum_i \lambda_i^2 \mathrm{Tr}\,\rho_0 E_i$ (at $\theta = 0$), which is exactly the left-hand side of the quantum inequality provided that

$$\varphi_{\rho_0}[B,C] = \tfrac{1}{2}\mathrm{Tr}\,\rho_0(BC + CB).$$

Generally, let us interpret the left-hand side as a sort of generalized variance of A. To do this it is useful to assume that

$$\varphi_\rho[B,B] = \mathrm{Tr}\,\rho B^2 \quad \text{if} \quad B\rho = \rho B. \tag{10.13}$$

However, in the non-commutative situation the statistical interpretation seems to be rather problematic and thus this quantity is called "quadratic cost functional."

The right-hand side of (10.12) is independent of the estimator and provides a lower bound for the quadratic cost. The denominator $\varphi_0[L,L]$ appears to be in the role of Fisher information here. It is called **quantum Fisher information** with respect to the cost function $\varphi_0[\cdot,\cdot]$. This quantity depends on the tangent of the curve ρ_θ. If the densities ρ_θ and the estimator A commute, then

$$L = \rho_0^{-1}\frac{d\rho_\theta}{d\theta} = \frac{d}{d\theta}\log\rho_\theta \quad \text{and} \quad \varphi_0[L,L] = \mathrm{Tr}\,\rho_0^{-1}\left(\frac{d\rho_\theta}{d\theta}\right)^2 = \mathrm{Tr}\,\rho_0\left(\rho_0^{-1}\frac{d\rho_\theta}{d\theta}\right)^2.$$
$$\tag{10.14}$$

The first formula justifies that L is called **logarithmic derivative**.

10.3 Quantum Fisher Information

A **coarse-graining** is an affine mapping sending density matrices into density matrices. Such a mapping extends to all matrices and provides a positivity and trace-preserving linear transformation. A common example of coarse-graining sends the density matrix ρ_{12} of a composite system $1 + 2$ into the (reduced) density matrix ρ_1 of component 1. There are several reasons to assume complete positivity about a coarse-graining and it is done so. Mathematically, coarse-graining is the same as state transformation in information channels. The terminology "coarse-graining" is used when the statistical aspects are focused on. Coarse-graining is the quantum analogue of a statistic.

Assume that $\rho_\theta = \rho + \theta B$ is a smooth curve of density matrices with tangent $B := \dot\rho$ at ρ. The quantum Fisher information $F_\rho(B)$ is an information quantity associated with the pair (ρ, B); it appeared in the Cramér–Rao inequality, and the classical Fisher information gives a bound for the variance of a locally unbiased estimator. Now let α be a coarse-graining. Then $\alpha(\rho_\theta)$ is another curve in the state space. Due to the linearity of α, the tangent at $\alpha(\rho_0)$ is $\alpha(B)$. As it is usual in statistics, information cannot be gained by coarse-graining; therefore it is expected that the Fisher information at the density matrix ρ_0 in the direction B must be larger than the Fisher information at $\alpha(\rho_0)$ in the direction $\alpha(B)$. This is the monotonicity property of the Fisher information under coarse-graining:

$$F_\rho(B) \geq F_{\alpha(\rho)}(\alpha(B)) \qquad (10.15)$$

Although we do not want to have a concrete formula for the quantum Fisher information, we require that this monotonicity condition must hold. Another requirement is that $F_\rho(B)$ should be quadratic in B; in other words, there exists a non-degenerate real bilinear form $\gamma_\rho(B,C)$ on the self-adjoint matrices such that

$$F_\rho(B) = \gamma_\rho(B,B). \qquad (10.16)$$

When ρ is regarded as a point of a manifold consisting of density matrices and B is considered as a tangent vector at the foot point ρ, the quadratic quantity $\gamma_\rho(B,B)$ may be regarded as a Riemannian metric on the manifold. This approach gives a geometric interpretation to the Fisher information.

The requirements (10.15) and (10.16) are strong enough to obtain a reasonable but still wide class of possible quantum Fisher informations.

We may assume that

$$\gamma_\rho(B,C) = \mathrm{Tr}\, B \mathbb{J}_\rho^{-1}(C) \qquad (10.17)$$

for an operator $\mathbb{J}_\rho$ acting on all matrices. (This formula expresses the inner product γ_ρ by means of the Hilbert–Schmidt inner product and the positive linear operator $\mathbb{J}_\rho$.) In terms of the operator $\mathbb{J}_\rho$ the monotonicity condition reads as

$$\alpha^* \mathbb{J}_{\alpha(\rho)}^{-1} \alpha \leq \mathbb{J}_\rho^{-1} \qquad (10.18)$$

for every coarse-graining α. (α^* stands for the adjoint of α with respect to the Hilbert–Schmidt product. Recall that α is completely positive and trace-preserving if and only if α^* is completely positive and unital.) On the other hand, the latter condition is equivalent to

$$\alpha \mathbb{J}_\rho \alpha^* \leq \mathbb{J}_{\alpha(\rho)}. \qquad (10.19)$$

It is interesting to observe the relevance of a certain quasi-entropy in the sense of (3.42):

$$\langle B\rho^{1/2}, f(\mathbb{L}_\rho \mathbb{R}_\rho^{-1}) B\rho^{1/2}\rangle = S_f^B(\rho\|\rho),$$

where the linear transformations $\mathbb{L}_\rho$ and $\mathbb{R}_\rho$ acting on matrices are the left and right multiplications, that is,

$$\mathbb{L}_\rho(X) = \rho X \qquad \text{and} \qquad \mathbb{R}_\rho(X) = X\rho.$$

When $f : \mathbb{R}^+ \to \mathbb{R}$ is operator monotone, then

$$\langle \alpha^*(B)\rho^{1/2}, f(\mathbb{L}_\rho \mathbb{R}_\rho^{-1}) \alpha^*(B)\rho^{1/2}\rangle \leq \langle B\alpha(\rho)^{1/2}, f(\mathbb{L}_{\alpha(\rho)} \mathbb{R}_{\alpha(\rho)}^{-1}) B\alpha(\rho)^{1/2}\rangle$$

due to the monotonicity of the quasi-entropy. If we set as

$$\mathbb{J}_\rho = \mathbb{R}_\rho^{1/2} f(\mathbb{L}_\rho \mathbb{R}_\rho^{-1}) \mathbb{R}_\rho^{1/2},$$

then (10.19) holds.

$$\varphi_\rho[B,B] := \operatorname{Tr} B\mathbb{J}_\rho(B) = \langle B\rho^{1/2}, f(\mathbb{L}_\rho\mathbb{R}_\rho^{-1})B\rho^{1/2}\rangle \qquad (10.20)$$

can be called **quadratic cost function**, and the corresponding monotone **quantum Fisher information**

$$\gamma_\rho(B,C) = \operatorname{Tr} B\mathbb{J}_\rho^{-1}(C) \qquad (10.21)$$

will be real for self-adjoint B and C if the function f satisfies the condition $f(t) = tf(t^{-1})$.

Example 10.11. In order to understand the action of the operator $\mathbb{J}_\rho$, assume that ρ is diagonal, $\rho = \sum_i p_i E_{ii}$. Then one can check that the matrix units E_{kl} are eigenvectors of $\mathbb{J}_\rho$, namely

$$\mathbb{J}_\rho(E_{kl}) = p_l f(p_k/p_l) E_{kl}.$$

The condition $f(t) = tf(t^{-1})$ gives that the eigenvectors E_{kl} and E_{lk} have the same eigenvalues. Therefore, the symmetrized matrix units $E_{kl} + E_{lk}$ and $iE_{kl} - iE_{lk}$ are eigenvectors as well.

Since

$$B = \sum_{k<l} \operatorname{Re} B_{kl}(E_{kl} + E_{lk}) + \sum_{k<l} \operatorname{Im} B_{kl}(iE_{kl} - iE_{lk}) + \sum_i B_{ii}E_{ii},$$

we have

$$\gamma_\rho(B,B) = 2\sum_{k<l} \frac{1}{p_k f(p_k/p_l)}|B_{kl}|^2 + \sum_i |B_{ii}|^2 \frac{1}{p_i}. \qquad (10.22)$$

In place of $2\sum_{k<l}$, we can write $\sum_{k\neq l}$. $\qquad\qquad\qquad\qquad\qquad\qquad\qquad\square$

Any monotone cost function has the property $\varphi_\rho[B,B] = \operatorname{Tr}\rho B^2$ for commuting ρ and B. The examples below show that it is not so generally.

Example 10.12. The analysis of operator monotone functions leads to the fact that among all monotone quantum Fisher informations there is a smallest one which corresponds to the (largest) function $f_m(t) = (1+t)/2$. In this case

$$F_\rho^{\min}(B) = \operatorname{Tr} BL = \operatorname{Tr}\rho L^2, \qquad \text{where} \qquad \rho L + L\rho = 2B. \qquad (10.23)$$

For the purpose of a quantum Cramér–Rao inequality the minimal quantity seems to be the best, since the inverse gives the largest lower bound. In fact, the matrix L has been used for a long time under the name of **symmetric logarithmic derivative**. In this example the quadratic cost function is

$$\varphi_\rho[B,C] = \tfrac{1}{2}\operatorname{Tr}\rho(BC + CB) \qquad (10.24)$$

and we have

$$\mathbb{J}_\rho(B) = \tfrac{1}{2}(\rho B + B\rho) \qquad \text{and} \qquad \mathbb{J}_\rho^{-1}(C) = 2\int_0^\infty e^{-t\rho} C e^{-t\rho}\, dt \qquad (10.25)$$

for the operator $\mathbb{J}_\rho$. Since $\mathbb{J}_\rho^{-1}$ is the smallest, $\mathbb{J}_\rho$ is the largest (among all possibilities).

There is a largest among all monotone quantum Fisher informations and this corresponds to the function $f_M(t) = 2t/(1+t)$. In this case

$$\mathbb{J}_\rho^{-1}(B) = \tfrac{1}{2}(\rho^{-1}B + B\rho^{-1}) \quad \text{and} \quad F_\rho^{\max}(B) = \mathrm{Tr}\,\rho^{-1}B^2. \tag{10.26}$$

It can be proved that the function

$$f_\alpha(t) = \alpha(1-\alpha)\frac{(t-1)^2}{(t^\alpha - 1)(t^{1-\alpha} - 1)} \tag{10.27}$$

is operator monotone for $\alpha \in (0,1)$ (see the Appendix, Example 11.20). F^α denotes the corresponding Fisher information metric. When $B = i[\rho,C]$ is orthogonal to the commutator of the foot point ρ in the tangent space, we have

$$F_\rho^\alpha(B) = \frac{1}{2\alpha(1-\alpha)}\mathrm{Tr}\,([\rho^\alpha, C][\rho^{1-\alpha}, C]). \tag{10.28}$$

Apart from a constant factor this expression is the **skew information** proposed by Wigner and Yanase some time ago [119]. In the limiting cases $\alpha \to 0$ or 1, we have

$$f_0(t) = \frac{1-t}{\log t}$$

and the corresponding quantum Fisher information

$$\gamma_\rho^0(B,C) = K_\rho(B,C) := \int_0^\infty \mathrm{Tr}\,B(\rho+t)^{-1}C(\rho+t)^{-1}\,dt \tag{10.29}$$

will be named here after Kubo and Mori. The **Kubo–Mori inner product** plays a role in quantum statistical mechanics. In this case $\mathbb{J}$ is the so-called **Kubo transform** $\mathbb{K}$ (and $\mathbb{J}^{-1}$ is the inverse Kubo transform $\mathbb{K}^{-1}$),

$$\mathbb{K}_\rho^{-1}(B) := \int_0^\infty (\rho+t)^{-1}B(\rho+t)^{-1}\,dt \quad \text{and} \quad \mathbb{K}_\rho(C) := \int_0^1 \rho^t C\rho^{1-t}\,dt. \tag{10.30}$$

Therefore the corresponding generalized variance is

$$\varphi_\rho[B,C] = \int_0^1 \mathrm{Tr}\,B\rho^t C\rho^{1-t}\,dt. \tag{10.31}$$

All pieces Fisher information discussed in this example are possible Riemannian metrics of manifolds of invertible density matrices. (Manifolds of pure states are rather different.) □

Fisher information appears not only as a Riemannian metric but as an information matrix as well. Let $\mathcal{M} := \{\rho_\theta : \theta \in G\}$ be a smooth m-dimensional manifold of invertible density matrices. The **quantum score operators** (or **logarithmic deriva-**

tives) are defined as

$$L_i(\theta) := \mathbb{J}_{\rho_\theta}^{-1}(\partial_{\theta_i}\rho_\theta) \qquad (1 \le i \le m) \tag{10.32}$$

and

$$Q_{ij}(\theta) := \mathrm{Tr}\, L_i(\theta)\mathbb{J}_{\rho_\theta}(L_j(\theta)) \qquad (1 \le i, j \le m) \tag{10.33}$$

is the **quantum Fisher information matrix**. This matrix depends on an operator monotone function which is involved in the super-operators $\mathbb{J}$. Historically the matrix Q determined by the symmetric logarithmic derivative (or the function $f_m(t) = (1+t)/2$) appeared first in the work of Helstrøm. Therefore, we call this **Helstrøm information matrix** and it will be denoted by $H(\theta)$.

The monotonicity of the Fisher information matrix is the manifestation of information loss under coarse-graining [98].

Theorem 10.3. *Fix an operator monotone function f to induce quantum Fisher information. Let α be a coarse-graining sending density matrices on the Hilbert space $\mathcal{H}_1$ into those acting on the Hilbert space $\mathcal{H}_2$ and let $\mathcal{M} := \{\rho_\theta : \theta \in G\}$ be a smooth m-dimensional manifold of invertible density matrices on $\mathcal{H}_1$. For the Fisher information matrix $Q^{(1)}(\theta)$ of $\mathcal{M}$ and for the Fisher information matrix $Q^{(2)}(\theta)$ of $\alpha(\mathcal{M}) := \{\alpha(\rho_\theta) : \theta \in G\}$, we have the monotonicity relation*

$$Q^{(2)}(\theta) \le Q^{(1)}(\theta). \tag{10.34}$$

(This is an inequality between $m \times m$ positive matrices.)

Proof. Set $B_i(\theta) := \partial_{\theta_i}\rho_\theta$. Then $\mathbb{J}_{\alpha(\rho_\theta)}^{-1}\alpha(B_i(\theta))$ is the score operator of $\alpha(\mathcal{M})$. Using (10.18), we have

$$\sum_{ij} Q_{ij}^{(2)}(\theta)a_i\overline{a_j} = \mathrm{Tr}\,\mathbb{J}_{\alpha(\rho_\theta)}^{-1}\alpha\Big(\sum_i a_iB_i(\theta)\Big)\alpha\Big(\sum_j \overline{a_j}B_j(\theta)\Big)$$

$$\le \mathrm{Tr}\,\mathbb{J}_{\rho_\theta}^{-1}\Big(\sum_i a_iB_i(\theta)\Big)\Big(\sum_j \overline{a_j}B_j(\theta)\Big)$$

$$= \sum_{ij} Q_{ij}^{(1)}(\theta)a_i\overline{a_j}$$

for any numbers a_i.   □

Assume that F_j are positive operators acting on a Hilbert space $\mathcal{H}_1$ on which the family $\mathcal{M} := \{\rho_\theta : \theta \in G\}$ is given. When $\sum_{j=1}^n F_j = I$, these operators determine a measurement. For any ρ_θ the formula

$$\alpha(\rho_\theta) := \mathrm{Diag}\,(\mathrm{Tr}\,\rho_\theta F_1, \dots, \mathrm{Tr}\,\rho_\theta F_n)$$

gives a diagonal density matrix. Since this family is commutative, all quantum Fisher informations coincide with the classical (10.14) and the classical Fisher information stands on the left-hand side of (10.34). We have

$$\mathbf{I}(\theta) \leq Q(\theta). \tag{10.35}$$

Combination of the classical Cramér–Rao inequality in Theorem 10.1 and (10.35) yields the **Helstrøm inequality**:

$$V(\theta) \geq H(\theta)^{-1}. \tag{10.36}$$

Example 10.13. In this example, let us investigate (10.35), which is equivalently written as

$$Q(\theta)^{-1/2}\mathbf{I}(\theta)Q(\theta)^{-1/2} \leq I_m.$$

Taking the trace, we have

$$\operatorname{Tr} Q(\theta)^{-1}\mathbf{I}(\theta) \leq m. \tag{10.37}$$

Assume that

$$\rho_\theta = \rho + \sum_k \theta_k B_k,$$

where $\operatorname{Tr} B_k = 0$ and the self-adjoint matrices B_k are pairwise orthogonal with respect to the inner product $(B,C) \mapsto \operatorname{Tr} B\mathbb{J}_\rho^{-1}(C)$.

The quantum Fisher information matrix

$$Q_{kl}(0) = \operatorname{Tr} B_k \mathbb{J}_\rho^{-1}(B_l)$$

is diagonal due to our assumption. Example 10.9 tells us about the classical Fisher information matrix:

$$\mathbf{I}_{kl}(0) = \sum_j \frac{\operatorname{Tr} B_k F_j \operatorname{Tr} B_l F_j}{\operatorname{Tr} \rho F_j}$$

Therefore,

$$\operatorname{Tr} Q(0)^{-1}\mathbf{I}(0) = \sum_k \frac{1}{\operatorname{Tr} B_k \mathbb{J}_\rho^{-1}(B_k)} \sum_j \frac{(\operatorname{Tr} B_k F_j)^2}{\operatorname{Tr} \rho F_j}$$

$$= \sum_j \frac{1}{\operatorname{Tr} \rho F_j} \sum_k \left(\operatorname{Tr} \frac{B_k}{\sqrt{\operatorname{Tr} B_k \mathbb{J}_\rho^{-1}(B_k)}} \mathbb{J}_\rho^{-1}(\mathbb{J}_\rho F_j) \right)^2.$$

We can estimate the second sum using the fact that

$$\frac{B_k}{\sqrt{\operatorname{Tr} B_k \mathbb{J}_\rho^{-1}(B_k)}}$$

is an orthonormal system and it remains so when ρ is added to it:

$$(\rho, B_k) = \operatorname{Tr} B_k \mathbb{J}_\rho^{-1}(\rho) = \operatorname{Tr} B_k = 0$$

and

$$(\rho, \rho) = \operatorname{Tr} \rho \mathbb{J}_\rho^{-1}(\rho) = \operatorname{Tr} \rho = 1.$$

Due to the Parseval inequality, we have

$$\left(\operatorname{Tr} \rho \mathbb{J}_\rho^{-1}(\mathbb{J}_\rho F_j)\right)^2 + \sum_k \left(\operatorname{Tr} \frac{B_k}{\sqrt{\operatorname{Tr} B_k \mathbb{J}_\rho^{-1}(B_k)}} \mathbb{J}_\rho^{-1}(\mathbb{J}_\rho F_j)\right)^2 \leq \operatorname{Tr}(\mathbb{J}_\rho F_j)\mathbb{J}_\rho^{-1}(\mathbb{J}_\rho F_j)$$

and

$$\operatorname{Tr} Q(0)^{-1}\mathbf{I}(0) \leq \sum_j \frac{1}{\operatorname{Tr} \rho F_j}\left(\operatorname{Tr}(\mathbb{J}_\rho F_j)F_j - (\operatorname{Tr} \rho F_j)^2\right)$$

$$= \sum_{j=1}^{n} \frac{\operatorname{Tr}(\mathbb{J}_\rho F_j)F_j}{\operatorname{Tr} \rho F_j} - 1 \leq n - 1$$

if we show that

$$\operatorname{Tr}(\mathbb{J}_\rho F_j)F_j \leq \operatorname{Tr} \rho F_j.$$

To see this we can use the fact that the left-hand side is a quadratic cost and it can be majorized by the largest one:

$$\operatorname{Tr}(\mathbb{J}_\rho F_j)F_j \leq \operatorname{Tr} \rho F_j^2 \leq \operatorname{Tr} \rho F_j,$$

because $F_j^2 \leq F_j$.

We obtained that

$$\operatorname{Tr} Q(\theta)^{-1}\mathbf{I}(\theta) \leq n - 1, \tag{10.38}$$

which can be compared with (10.37). This bound can be smaller than the general one. The assumption on B_k's is not very essential, since the orthogonality can be reached by reparameterization.   □

Let $\mathcal{M} := \{\rho_\theta : \theta \in G\}$ be a smooth m-dimensional manifold and assume that a collection $A = (A_1, \ldots, A_m)$ of self-adjoint matrices is used to estimate the true value of θ.

Given an operator $\mathbb{J}$ we have the corresponding cost function $\varphi_\theta \equiv \varphi_{\rho_\theta}$ for every θ and the **cost matrix** of the estimator A is a positive definite matrix, defined by $\varphi_\theta[A]_{ij} = \varphi_\theta[A_i, A_j]$. The **bias** of the estimator is

$$b(\theta) = \left(b_1(\theta), b_2(\theta), \ldots, b_m(\theta)\right)$$
$$:= \left(\operatorname{Tr} \rho_\theta(A_1 - \theta_1), \operatorname{Tr} \rho_\theta(A_2 - \theta_2), \ldots, \operatorname{Tr} \rho_\theta(A_m - \theta_m)\right).$$

From the bias vector we can form a **bias matrix**

$$B_{ij}(\theta) := \partial_{\theta_i} b_j(\theta) \qquad (1 \leq i, j \leq m).$$

For a locally unbiased estimator at θ_0, we have $B(\theta_0) = 0$.

The next result is the quantum Cramér–Rao inequality for a biased estimate.

Theorem 10.4. *Let $A = (A_1, \ldots, A_m)$ be an estimator of θ. Then for the above-defined quantities the inequality*

$$\varphi_\theta[A] \geq (I + B(\theta)) Q(\theta)^{-1} (I + B(\theta)^*)$$

holds in the sense of the order on positive semidefinite matrices. (Here I denotes the identity operator.)

Proof. Let us use the block-matrix method. Let X and Y be $m \times m$ matrices with $n \times n$ entries and assume that all entries of Y are constant multiples of the unit matrix. (A_i and L_i are $n \times n$ matrices.) If α is a completely positive mapping on $n \times n$ matrices, then $\tilde{\alpha} := \mathrm{Diag}(\alpha, \ldots, \alpha)$ is a positive mapping on block matrices and $\tilde{\alpha}(YX) = Y\tilde{\alpha}(X)$. This implies that $\mathrm{Tr}\, X\alpha(X^*)Y \geq 0$ when Y is positive. Therefore the $m \times m$ ordinary matrix M which has ij entry

$$\mathrm{Tr}\,(X\tilde{\alpha}(X^*))_{ij}$$

is positive. In the sequel let us restrict ourselves to $m = 2$ for the sake of simplicity and apply the above fact to the case

$$X = \begin{bmatrix} A_1 & 0 & 0 & 0 \\ A_2 & 0 & 0 & 0 \\ L_1(\theta) & 0 & 0 & 0 \\ L_2(\theta) & 0 & 0 & 0 \end{bmatrix} \quad \text{and} \quad \alpha = \mathbb{J}_{\rho_\theta}.$$

Then we have

$$M = \begin{bmatrix} \mathrm{Tr}\, A_1 \mathbb{J}_\rho(A_1) & \mathrm{Tr}\, A_1 \mathbb{J}_\rho(A_2) & \mathrm{Tr}\, A_1 \mathbb{J}_\rho(L_1) & \mathrm{Tr}\, A_1 \mathbb{J}_\rho(L_2) \\ \mathrm{Tr}\, A_2 \mathbb{J}_\rho(A_1) & \mathrm{Tr}\, A_2 \mathbb{J}_\rho(A_2) & \mathrm{Tr}\, A_2 \mathbb{J}_\rho(L_1) & \mathrm{Tr}\, A_2 \mathbb{J}_\rho(L_2) \\ \mathrm{Tr}\, L_1 \mathbb{J}_\rho(A_1) & \mathrm{Tr}\, L_1 \mathbb{J}_\rho(A_2) & \mathrm{Tr}\, L_1 \mathbb{J}_\rho(L_1) & \mathrm{Tr}\, L_1 \mathbb{J}_\rho(L_2) \\ \mathrm{Tr}\, L_2 \mathbb{J}_\rho(A_1) & \mathrm{Tr}\, L_2 \mathbb{J}_\rho(A_2) & \mathrm{Tr}\, L_2 \mathbb{J}_\rho(L_1) & \mathrm{Tr}\, L_2 \mathbb{J}_\rho(L_2) \end{bmatrix} \geq 0$$

Now we can rewrite the matrix M in terms of the matrices involved in our Cramér–Rao inequality. The 2×2 block M_{11} is the generalized covariance, M_{22} is the Fisher information matrix and M_{12} is easily expressed as $I + B$. We will get

$$M = \begin{bmatrix} \varphi_\theta[A_1, A_1] & \varphi_\theta[A_1, A_2] & 1 + B_{11}(\theta) & B_{12}(\theta) \\ \varphi_\theta[A_2, A_1] & \varphi_\theta[A_2, A_2] & B_{21}(\theta) & 1 + B_{22}(\theta) \\ 1 + B_{11}(\theta) & B_{21}(\theta) & \varphi_\theta[L_1, L_1] & \varphi_\theta[L_1, L_2] \\ B_{12}(\theta) & 1 + B_{22}(\theta) & \varphi_\theta[L_2, L_1] & \varphi_\theta[L_2, L_2] \end{bmatrix} \geq 0$$

The positivity of a block-matrix

$$M = \begin{bmatrix} M_1 & C \\ C^* & M_2 \end{bmatrix} = \begin{bmatrix} \varphi_\rho[A] & I + B \\ I + B^* & J(\theta) \end{bmatrix}$$

implies $M_1 \geq CM_2^{-1}C^*$, which reveals exactly the statement of the theorem. (Concerning positive block-matrices, see the Appendix.)   □

10.4 Contrast Functionals

Let M_Θ be a smooth manifold of density matrices. The following construction is motivated by classical statistics. Suppose that a positive functional $d(\rho_1, \rho_2)$ of two variables is given on the manifold. In many cases one can get a Riemannian metric by differentiation:

$$g_{ij}(\theta) = \frac{\partial^2}{\partial \theta_i \partial \theta_j'} d(\rho_\theta, \rho_{\theta'}) \Big|_{\theta = \theta'}, \qquad (\theta \in \Theta).$$

To be more precise the positive smooth functional $d(\cdot, \cdot)$ is called a **contrast functional** if $d(\rho_1, \rho_2) = 0$ implies $\rho_1 = \rho_2$.

Following the work of Csiszár in classical information theory, Petz introduced a family of information quantities parameterized by a function $F : \mathbb{R}^+ \to \mathbb{R}$

$$S_F(\rho_1, \rho_2) = \langle \rho_1^{1/2}, F(\Delta(\rho_2/\rho_1)) \rho_1^{1/2} \rangle, \tag{10.39}$$

see (3.23), F is written in place of f. ($\Delta(\rho_2/\rho_1) := L_{\rho_2} R_{\rho_1}^{-1}$ is the relative modular operator of the two densities.) When F is operator monotone decreasing, this quasi-entropy possesses good properties; for example, it is a contrast functional in the above sense if F is not linear and $F(1) = 0$. In particular, for

$$F_\alpha(t) = \frac{1}{\alpha(1-\alpha)} \left(1 - t^\alpha\right)$$

we have

$$S_\alpha(\rho_1, \rho_2) = \frac{1}{\alpha(1-\alpha)} \mathrm{Tr}\, (I - \rho_2^\alpha \rho_1^{-\alpha}) \rho_1 \tag{10.40}$$

The differentiation is

$$\frac{\partial^2}{\partial t \partial u} S_\alpha(\rho + tB, \rho + uC) = -\frac{1}{\alpha(1-\alpha)} \frac{\partial^2}{\partial t \partial u} \mathrm{Tr}\, (\rho + tB)^{1-\alpha} (\rho + uC)^\alpha =: K_\rho^\alpha(B, C)$$

at $t = u = 0$ in the affine parameterization. The tangent space at ρ is decomposed into two subspaces: the first consists of self-adjoint matrices commuting with ρ and the second is $\{i(D\rho - \rho D) : D = D^*\}$, the set of commutators. The decomposition is essential both from the viewpoint of differential geometry and from the point of view of differentiation (see Example 11.8). If B and C commute with ρ, then

$$K_\rho^\alpha(B, C) = \mathrm{Tr}\, \rho^{-1} BC$$

is independent of α and it is the classical Fischer information (in matrix form). If $B = \mathrm{i}[D_B,\rho]$ and $C = \mathrm{i}[D_C,\rho]$, then

$$K_\rho^\alpha(B,C) = Tr[\rho^{1-\alpha},D_B][\rho^\alpha,D_C].$$

This is related to **skew information** (10.28).

Ruskai and Lesniewski discovered that all pieces of monotone Fisher information are obtained from a quasi-entropy as contrast functional [70]. The relation of the function F in (10.39) to the function f in Theorem 10.3 is

$$\frac{1}{f(t)} = \frac{F(t)+tF(t^{-1})}{(t-1)^2}. \tag{10.41}$$

10.5 Notes

The quantum analogue of the Cramér–Rao inequality was discovered immediately after the foundation of mathematical quantum estimation theory in the 1960s (see the book [50] of Helstrom, or the book [56] of Holevo for a rigorous summary of the subject).

The monotone Riemannian metrics on density matrices were studied by Petz and he proved that such metrics can be given by an operator monotone function [93, 94]. Theorem 10.3 appeared in [98]. However, particular cases of the monotonicity are already in the earlier literature: [93] treated the case of the Kubo–Mori inner product and [22] considered the Helstrøm information matrix and measurement in the role of coarse-graining. Example 10.13 is from [42].

The Kubo–Mori inner product plays a role in quantum statistical mechanics (see [40], for example).

For the role of contrast functionals in classical estimation, see [37]. We can note that a contrast functional is a particular example of yokes (cf. [11]). The Ricmannian geometry of Fischer information is the subject of the book [5]. The differential geometries induced by different contrast functionals have not been studied in detail in the quantum case. For the contrast functional

$$d(\rho_1,\rho_2) = 1 - \inf\{Tr\rho_1^\alpha\rho_2^{1-\alpha} : 0 < \alpha < 1\}$$

the corresponding inner product was computed in [10].

The efficiency of a state estimation can be measured by the **mean quadratic error matrix**. The smaller the matrix is, the better is the efficiency. Since matrices are typically not comparable, a possibility is to minimize $Tr\,GM$, where M is the error matrix and G is a weight matrix. The efficiency of the state estimation can be increased by an **adaptive** attitude: Some measurements are made, a rough estimate is obtained and the refining measurements depend on the rough estimate [42, 46].

10.6 Exercises

1. Is the estimate Φ in Example 10.4 unbiased?
2. Show that the mean quadratic error of the estimate (10.4) is

$$V_n(\theta) = \frac{1}{n} \begin{bmatrix} 1 - \theta_1^2 & 0 & 0 \\ 0 & 1 - \theta_2^2 & 0 \\ 0 & 0 & 1 - \theta_3^2 \end{bmatrix}. \tag{10.42}$$

3. Calculate the classical Fisher information matrix $I(\theta)$ for Example 10.1.
4. Let $B(t) := e^{-t\rho} B e^{-t\rho}$. Use the fact

$$\frac{d}{dt} B(t) = -\rho B(t) - B(t)\rho$$

to show that

$$L = 2 \int_0^\infty e^{-t\rho} B e^{-t\rho} \, dt$$

is the solution of the equation $\rho L + L\rho = 2B$.
5. In the setting of Example 10.4, use the estimate

$$\Phi(i) = \frac{1}{2}(I + a_i \cdot \sigma).$$

Show that Φ is unbiased.
6. Let ρ be a state and the mapping $\mathbb{K}_\rho(C) = \int_0^1 \rho^t C \rho^{1-t} \, dt$ be defined for self-adjoint matrices. Show that

$$\mathbb{K}_\rho^{-1}(B) = \int_0^\infty (\rho + t)^{-1} B(\rho + t)^{-1} \, dt.$$

Show that

$$\mathbb{K}_\rho\{C : \mathrm{Tr}\,\rho C = 0\} = \{B : \mathrm{Tr}\,B = 0\}.$$

7. Show that there exists no observable which is an unbiased estimator for the family in Example 10.10.

Chapter 11
Appendix: Auxiliary Linear and Convex Analysis

Quantum information theory is an interdisciplinary field: The physical background is provided by quantum mechanics, the questions are given in the language of information theory but the mathematical technicalities are provided by linear analysis. Since the Hilbert spaces of the quantum mechanical description is finite dimensional in many cases, matrix theory can be the substitute of the much more delicate operator theory and matrix manipulations play an important role.

11.1 Hilbert Spaces and Their Operators

Let $\mathscr{H}$ be a complex vector space. A functional $\langle \cdot, \cdot \rangle : \mathscr{H} \times \mathscr{H} \to \mathbb{C}$ of two variables is called **inner product** if

(1) $\langle x+y, z \rangle = \langle x, z \rangle + \langle y, z \rangle$ $(x, y, z \in \mathscr{H})$,

(2) $\langle \lambda x, y \rangle = \overline{\lambda} \langle x, y \rangle$ $(\lambda \in \mathbb{C}, x, y \in \mathscr{H})$,

(3) $\langle x, y \rangle = \overline{\langle y, x \rangle}$ $(x, y \subset \mathscr{H})$,

(4) $\langle x, x \rangle \geq 0$ for every $x \in \mathscr{H}$ and $\langle x, x \rangle = 0$ only for $x = 0$.

These conditions imply the **Schwarz inequality**

$$|\langle x, y \rangle|^2 \leq \langle x, x \rangle \langle y, y \rangle. \tag{11.1}$$

The inner product determines a **norm**

$$\|x\| := \sqrt{\langle x, x \rangle} \tag{11.2}$$

which has the properties

$$\|x+y\| \leq \|x\| + \|y\| \qquad \text{and} \qquad |\langle x, y \rangle| \leq \|x\| \cdot \|y\|.$$

D. Petz, *Appendix: Auxiliary Linear and Convex Analysis*. In: D. Petz, Quantum Information Theory and Quantum Statistics, Theoretical and Mathematical Physics, pp. 165–203 (2008)
DOI 10.1007/978-3-540-74636-2_11   © Springer-Verlag Berlin Heidelberg 2008

$\|x\|$ is interpreted as the length of the vector x. A further requirement in the definition of a Hilbert space is that every Cauchy sequence must be convergent, that is, the space is complete.

If $\langle x, y \rangle = 0$ for vectors x and y of a Hilbert space, then x and y are called **orthogonal**, in notation $x \perp y$. When $H \subset \mathscr{H}$, then $H^{\perp} := \{x \in \mathscr{H} : x \perp h \text{ for every } h \in H\}$. For any subset $H \subset \mathscr{H}$ the orthogonal complement $H^{\perp}$ is a closed subspace.

A family $\{e_i\}$ of vectors is called **orthonormal** if $\langle e_i, e_i \rangle = 1$ and $\langle e_i, e_j \rangle = 0$ if $i \neq j$. A maximal orthonormal system is called **basis**. The cardinality of a basis is called the dimension of the Hilbert space. (The cardinality of any two bases is the same.)

Theorem 11.1. *Let $e_1, e_2, \ldots$ be a basis in a Hilbert space $\mathscr{H}$. Then for any vector $x \in \mathscr{H}$ the expansion*

$$x = \sum_n \langle e_n, x \rangle e_n$$

holds. Moreover,

$$\|x\|^2 = \sum_n |\langle e_n, x \rangle|^2$$

Theorem 11.2. (Projection theorem) *Let $\mathscr{M}$ be a closed subspace of a Hilbert space $\mathscr{H}$. Any vector $x \in \mathscr{H}$ can be written in a unique way in the form $x = x_0 + y$, where $x_0 \in \mathscr{M}$ and $y \perp \mathscr{M}$.*

The mapping $P : x \mapsto x_0$ defined in the context of the previous theorem is called **orthogonal projection** onto the subspace $\mathscr{M}$. This mapping is linear:

$$P(\lambda x + \mu y) = \lambda P x + \mu P y.$$

Moreover, $P^2 = P$.

Let $A : \mathscr{H} \to \mathscr{H}$ be a linear mapping and $e_1, e_2, \ldots, e_n$ be a basis in the Hilbert space $\mathscr{H}$. The mapping A is determined by the vectors $A e_k$, $k = 1, 2, \ldots, n$. Furthermore, the vector $A e_k$ is determined by its coordinates:

$$A e_k = c_{1,k} e_1 + c_{2,k} e_2 + \cdots + c_{n,k} e_n.$$

The numbers $c_{i,j}$ form an $n \times n$ matrix; it is called the **matrix** of the linear transformation A in the basis $e_1, e_2, \ldots, e_n$.

The **norm** of a linear operator $A : \mathscr{H} \to \mathscr{K}$ is defined as

$$\|A\| := \sup\{\|Ax\| : x \in \mathscr{H}, \|x\| = 1\}.$$

The linear oparator A is called **bounded** if $\|A\|$ is finite. The set of all bounded operators $\mathscr{H} \to \mathscr{H}$ is denoted by $B(\mathscr{H})$.

Let $\mathscr{H}$ and $\mathscr{K}$ be Hilbert spaces. If $T : \mathscr{H} \to \mathscr{K}$ is a bounded linear operator, then its **adjoint** $T^* : \mathscr{K} \to \mathscr{H}$ is determined by the formula

$$\langle x, Ty \rangle_{\mathscr{K}} = \langle T^*x, y \rangle_{\mathscr{H}} \qquad (x \in \mathscr{K}, y \in \mathscr{H}). \tag{11.3}$$

The operator $T \in B(\mathscr{H})$ is called **self-adjoint** if $T^* = T$. The operator T is self-adjoint if and only if $\langle x, Tx \rangle$ is a real number for every vector $x \in \mathscr{H}$.

Theorem 11.3. *The properties of the adjoint are as follows:*

(1) $(A+B)^* = A^* + B^*$, $(\lambda A)^* = \overline{\lambda} A^*$ $\qquad (\lambda \in \mathbb{C})$,

(2) $(A^*)^* = A$, $(AB)^* = B^*A^*$,

(3) $(A^{-1})^* = (A^*)^{-1}$ *if A is invertible,*

(4) $\|A\| = \|A^*\|$, $\|A^*A\| = \|A\|^2$.

Example 11.1. Let $A : \mathscr{H} \to \mathscr{H}$ be a linear mapping and $e_1, e_2, \ldots, e_n$ be a basis in the Hilbert space $\mathscr{H}$. The (i, j) element of the matrix of A is $\langle e_i, Ae_j \rangle$. Since

$$\langle e_i, Ae_j \rangle = \overline{\langle e_j, A^*e_i \rangle},$$

this is the complex conjugate of the (j, i) element of the matrix of A^*. $\qquad\qquad \square$

The operators we need are mostly linear, but sometimes **conjugate-linear** operators appear. $\Lambda : \mathscr{H} \to \mathscr{K}$ is conjugate-linear if

$$\Lambda(\lambda x + \mu y) = \overline{\lambda} x + \overline{\mu} y$$

for any complex numbers λ and μ and for any vectors $x, y \in \mathscr{H}$. The adjoint Λ^* of the conjugate-linear operararator Λ is determined by the equation

$$\langle x, \Lambda y \rangle_{\mathscr{K}} = \langle y, \Lambda^* x \rangle_{\mathscr{H}} \qquad (x \in \mathscr{K}, y \in \mathscr{H}). \tag{11.4}$$

11.2 Positive Operators and Matrices

Let $\mathscr{H}$ be a Hilbert space and $T : \mathscr{H} \to \mathscr{H}$ be a bounded linear operator. T is called **positive** (or positive semidefinite) if $\langle x, Tx \rangle \geq 0$ for every vector $x \in \mathscr{H}$, in notation $T \geq 0$. It follows from the definition that a positive operator is self-adjoint. Moreover, if T_1 and T_2 are positive operators, then $T_1 + T_2$ is positive as well.

Theorem 11.4. *Let $T \in B(\mathscr{H})$ be an operator. The following conditions are equivalent.*

(1) T is positive.

(2) $T = T^*$ and the spectrum of T lies in $\mathbb{R}^+$.

(3) T is of the form A^*A for some operator $A \in B(\mathscr{H})$.

Example 11.2. Let T be positive operator acting on a finite-dimensional Hilbert space such that $\|T\| \leq 1$. Let us show that there is unitary operator U such that

$$T = \frac{1}{2}(U + U^*).$$

We can choose an orthonormal basis $e_1, e_2, \ldots, e_n$ consisting of eigenvectors of T and in this basis the matrix of T is diagonal, say, $\text{Diag}(t_1, t_2, \ldots, t_n)$, $0 \leq t_j \leq 1$ from the positivity. For any $1 \leq j \leq n$ we can find a real number θ_j such that

$$t_j = \frac{1}{2}(e^{i\theta_j} + e^{-i\theta_j}).$$

Then the unitary operator U with matrix $\text{Diag}(\exp(i\theta_1)), \ldots, \exp(i\theta_n))$ will have the desired property. $\square$

If T acts on a finite-dimensional Hilbert space which has an orthonormal basis $e_1, e_2, \ldots, e_n$, then T is uniquely determined by its matrix

$$[\langle e_i, Te_j \rangle]_{i,j=1}^n.$$

T is positive if and only if its matrix is positive (semidefinite).

Example 11.3. Let

$$A = \begin{bmatrix} \lambda_1 & \lambda_2 & \ldots & \lambda_n \\ 0 & 0 & \ldots & 0 \\ \vdots & \vdots & \ddots & \vdots \\ 0 & 0 & \ldots & 0 \end{bmatrix}.$$

Then

$$[A^*A]_{i,j} = \overline{\lambda}_i \lambda_j \qquad (1 \leq i, j \leq n)$$

and this matrix is positive. Every positive matrix is the sum of matrices of this form. (The minimum number of the summands is the rank of the matrix.) $\square$

Theorem 11.5. *Let $T \in B(\mathcal{H})$ be a self-adjoint operator and $e_1, e_2, \ldots, e_n$ be a basis in the Hilbert space $\mathcal{H}$. T is positive if and only if for any $1 \leq k \leq n$ the determinant of the $k \times k$ matrix*

$$[\langle e_i, Te_j \rangle]_{ij=1}^k$$

is positive (that is, ≥ 0).

For a 2×2 matrix, it is very easy to check the positivity:

$$\begin{bmatrix} a & b \\ \overline{b} & c \end{bmatrix} \geq 0 \quad \text{if} \quad a \geq 0 \quad \text{and} \quad b\overline{b} \leq ac. \tag{11.5}$$

If the entries are $n \times n$ matrices, then the condition for positivity is similar but it is a bit more complicated. Matrices with matrix entries are called **block-matrices**.

Theorem 11.6. *The self-adjoint block-matrix*

$$\begin{bmatrix} A & B \\ B^* & C \end{bmatrix}$$

is positive if and only if $A, C \geq 0$ and there exists an operator X such that $\|X\| \leq 1$ and $B = C^{1/2} X A^{1/2}$. When A is invertible, then this condition is equivalent to

$$BA^{-1}B^* \leq C.$$

For an invertible A, we have the so-called **Schur factorization**

$$\begin{bmatrix} A & B \\ B^* & C \end{bmatrix} = \begin{bmatrix} I & 0 \\ B^*A^{-1} & I \end{bmatrix} \cdot \begin{bmatrix} A & 0 \\ 0 & C - B^*A^{-1}B \end{bmatrix} \cdot \begin{bmatrix} I & A^{-1}B \\ 0 & I \end{bmatrix}. \tag{11.6}$$

Since

$$\begin{bmatrix} I & 0 \\ B^*A^{-1} & I \end{bmatrix}^{-1} = \begin{bmatrix} I & 0 \\ -B^*A^{-1} & I \end{bmatrix}$$

is invertible, the positivity of the left-hand side of (11.6) is equivalent to the positivity of the middle factor of the right-hand side.

Theorem 11.7. (Schur) *Let A and B be positive $n \times n$ matrices. Then*

$$C_{ij} = A_{ij}B_{ij} \qquad (1 \leq i, j \leq n)$$

determines a positive matrix.

Proof. If $A_{ij} = \overline{\lambda}_i \lambda_j$ and $B_{ij} = \overline{\mu}_i \mu_j$, then $C_{ij} = \overline{\lambda_i \mu_i} \lambda_j \mu_j$ and C is positive as per Example 11.3. The general case is reduced to this one.   □

The matrix C of the previous theorem is called the **Hadamard** (or Schur) **product** of the matrices A and B. In notation, $C = A \circ B$.

Let $A, B \in B(\mathcal{H})$ be self-adjoint operators. $A \leq B$ if $B - A$ is positive. The inequality $A \leq B$ implies $XAX^* \leq XBX^*$ for every operator X.

Example 11.4. Let $\mathcal{K}$ be a closed subspace of a Hilbert space $\mathcal{H}$. Any vector $x \in \mathcal{H}$ can be written in the form $x_0 + x_1$, where $x_0 \in \mathcal{K}$ and $x_1 \perp \mathcal{K}$. The linear mapping $P : x \mapsto x_0$ is called (orthogonal) **projection** onto $\mathcal{K}$. The orthogonal projection P has the properties $P = P^2 = P^*$. If an operator $P \in B(\mathcal{H})$ has the property $P = P^2 = P^*$, then it is a (orthogonal) projection (onto its range).

If P and Q are projections then the relation $P \leq Q$ means that the range of P is contained in the range of Q. An equivalent algebraic formulation is $PQ = P$.

If P is a projection, then $I - P$ is a projection as well and it is often denoted by $P^\perp$, since the range of $I - P$ is the orthogonal complement of the range of P.   □

Example 11.5. Let X be an operator on a Hilbert space $\mathscr{H}$ and let P be an orthogonal projection. The Hilbert space $\mathscr{H}$ is decomposed as the direct sum of the ranges of P and $P^\perp$, so X can be written in a block-matrix form

$$X = \begin{bmatrix} PXP & P^\perp XP \\ PXP^\perp & P^\perp XP^\perp \end{bmatrix}.$$

If X is invertible, then its inverse has a similar form:

$$X^{-1} = \begin{bmatrix} PX^{-1}P & P^\perp X^{-1}P \\ PX^{-1}P^\perp & P^\perp X^{-1}P^\perp \end{bmatrix}.$$

Formula (11.77) contains the entries of X^{-1} expressed by the entries of X. In particular,

$$PX^{-1}P = (PXP)^{-1} + \text{another term}.$$

If X is positive, then the other term is positive as well, and we can conclude

$$PX^{-1}P \geq (PXP)^{-1}. \tag{11.7}$$

(On the right-hand side the inverse is in the range space of P, or equivalently it is the solution of the equations $YP^\perp = 0$ and $Y(PXP) = P$.) $\qquad\square$

Theorem 11.8. *Let A be a positive $n \times n$ block-matrix with $k \times k$ entries. Then A is the sum of block-matrices B of the form $[B]_{ij} = X_i^* X_j$ for some $k \times k$ matrices $X_1, X_2, \ldots, X_n$.*

Proof. A can be written as C^*C for some

$$C = \begin{bmatrix} C_{11} & C_{12} & \cdots & C_{1n} \\ C_{21} & C_{22} & \cdots & C_{2n} \\ \vdots & \vdots & \ddots & \vdots \\ C_{n1} & C_{n2} & \cdots & C_{nn} \end{bmatrix}.$$

Let B_i be the block-matrix such that its ith raw is the same as in C and all other elements are 0. Then $C = B_1 + B_2 + \cdots + B_n$ and for $t \neq i$ we have $B_t^* B_i = 0$. Therefore,

$$A = (B_1 + B_2 + \cdots + B_n)^* (B_1 + B_2 + \cdots + B_n) = B_1^* B_1 + B_2^* B_2 + \cdots + B_n^* B_n.$$

The (i, j) entry of $B_t^* B_t$ is $C_{ti}^* C_{tj}$, hence this matrix is of the required form. $\qquad\square$

11.3 Functional Calculus for Matrices

Let $A \in M_n(\mathbb{C})$ and $p(x) := \sum_i c_i x^i$ be a polynomial. It is quite obvious that by $p(A)$ one means the matrix $\sum_i c_i A^i$. The **functional calculus** can be extended to other functions $f : \mathbb{C} \to \mathbb{C}$. When

$$A = \text{Diag}(\lambda_1, \lambda_2, \ldots, \lambda_n)$$

is diagonal and the function f is defined on the eigenvalues of A, then

$$f(A) = \text{Diag}(f(\lambda_1), f(\lambda_2), \ldots, f(\lambda_n)).$$

Assume now that A is diagonalizable, that is,

$$A = S\,\text{Diag}(\lambda_1, \lambda_2, \ldots, \lambda_n)S^{-1}$$

with an invertible matrix S. Then $f(A)$ is defined as

$$S\,\text{Diag}(f(\lambda_1), f(\lambda_2), \ldots, f(\lambda_n))S^{-1}$$

when the complex-valued function f is defined on the set of eigenvalues of A. Remember that self-adjoint matrices are diagonalizable and they have a spectral decomposition. Let $A = \sum_i \lambda_i P_i$ be the spectral decomposition of the self-adjoint $A \in M_n(\mathbb{C})$. (λ_i are the different eigenvalues and P_i are the corresponding eigenprojections, the rank of P_i is the multiplicity of λ_i.) Then

$$f(A) = \sum_i f(\lambda_i)P_i. \tag{11.8}$$

Usually it is assumed that f is continuous or even smooth on an interval containing the eigenvalues of A.

Let f be holomorphic inside and on a positively oriented simple contour Γ in the complex plane and let A be an $n \times n$ matrix such that its eigenvalues are the inside of Γ. Then

$$f(A) := \frac{1}{2\pi i} \int_\Gamma f(z)(zI - A)^{-1} dz \tag{11.9}$$

is defined by a contour integral. When A is self-adjoint, then (11.8) makes sense and it is an exercise to show that it gives the same result as (11.9).

Theorem 11.9. *Let $A, B \in M_n(\mathbb{C})$ be self-adjoint matrices and $t \in \mathbb{R}$. Assume that $f : (\alpha, \beta) \to \mathbb{R}$ is a continuously differentiable function defined on an interval and assume that the eigenvalues of $A + tB$ are in (α, β) for small $t - t_0$. Then*

$$\frac{d}{dt}\text{Tr}\,f(A + tB)\Big|_{t=t_0} = \text{Tr}\,(Bf'(A + t_0 B)),$$

Proof. One can verify the formula for a polynomial f by an easy direct computation and the argument can be extended to a more general f by means of polynomial approximation. □

Example 11.6. Let $f : (\alpha, \beta) \to \mathbb{R}$ is a continuous increasing function and assume that the spectrum of the self-adjoint matrices B and C lies in (α, β). Let us use the previous theorem to show that

$$A \le C \quad \text{implies} \quad \text{Tr}\,f(A) \le \text{Tr}\,f(C). \tag{11.10}$$

We may assume that f is smooth and it is enough to show that the derivative of $\operatorname{Tr} f(A+tB)$ is positive when $B \geq 0$. The derivative is $\operatorname{Tr}(Bf'(A+tB))$ and this is the trace of the product of two positive operators. Therefore, it is positive.   □

Example 11.7. For a holomorphic function f, we can compute the derivative of $f(A+tB)$ on the basis of (11.9), where Γ is a positively oriented simple contour satisfying the properties required above. The derivation is reduced to the differentiation of the resolvent $(zI - (A+tB))^{-1}$ and we obtain

$$\frac{d}{dt}f(A+tB)\Big|_{t=0} = \frac{1}{2\pi i}\int_\Gamma f(z)(zI-A)^{-1}B(zI-A)^{-1}\,dz. \tag{11.11}$$

When A is self-adjoint, then it is not a restriction to assume that it is diagonal, $A = \operatorname{Diag}(t_1, t_2, \ldots, t_n)$, and we shall compute the entries of the matrix (11.11). So

$$\left(\frac{d}{dt}f(A+tB)\Big|_{t=0}\right)_{ij} = \frac{f(t_i)-f(t_j)}{t_i-t_j} \times B_{ij}, \tag{11.12}$$

in other words, the derivative is the **Hadamard** (or entrywise) **product** of the divided difference matrix $[a_{ij}]_{i,j=1}^n$ defined as

$$a_{ij} = \begin{cases} \dfrac{f(t_i)-f(t_j)}{t_i-t_j} & \text{if } t_i - t_j \neq 0, \\[2ex] f'(t_i) & \text{if } t_i - t_j = 0, \end{cases} \tag{11.13}$$

and the matrix B.   □

Example 11.8. Next let us restrict ourselves to the self-adjoint case $A, B \in M_n(\mathbb{C})^{sa}$ in the analysis of (11.11).

The space $M_n(\mathbb{C})^{sa}$ can be decomposed as $\mathcal{M}_A \oplus \mathcal{M}_A^\perp$, where $\mathcal{M}_A := \{C \in M_n(\mathbb{C})^{sa} : CA = AC\}$ is the commutant of A and $\mathcal{M}_A^\perp$ is its orthogonal complement. When the operator $\mathcal{L}_A : X \mapsto i(AX - XA) \equiv i[A, X]$ is considered, $\mathcal{M}_A$ is exactly the kernel of $\mathcal{L}_A$, while $\mathcal{M}_A^\perp$ is its range.

When $B \in \mathcal{M}_A$, then

$$\frac{1}{2\pi i}\int_\Gamma f(z)(zI-A)^{-1}B(zI-A)^{-1}\,dz = \frac{B}{2\pi i}\int_\Gamma f(z)(zI-A)^{-2}\,dz = Bf'(A)$$

and we have

$$\frac{d}{dt}f(A+tB)\Big|_{t=0} = Bf'(A). \tag{11.14}$$

When $B = i[A, X] \in \mathcal{M}_A^\perp$, then we can use the identity

$$(zI-A)^{-1}[A, X](zI-A)^{-1} = [(zI-A)^{-1}, X]$$

and we can conclude

$$\frac{d}{dt}f(A+ti[A,X])\Big|_{t=0} = i[f(A),X]. \qquad (11.15)$$

To compute the derivative in an arbitrary direction B we should decompose B as $B_1 \oplus B_2$ with $B_1 \in \mathcal{M}_A$ and $B_2 \in \mathcal{M}_A^\perp$. Then

$$\frac{d}{dt}f(A+tB)\Big|_{t=0} = B_1 f'(A) + i[f(A),X], \qquad (11.16)$$

where X is the solution of the equation $B_2 = i[A,X]$. □

Let $J \subset \mathbb{R}$ be an interval. A function $f : J \to \mathbb{R}$ is said to be **convex** if

$$f(ta+(1-t)b) \le tf(a)+(1-t)f(b) \qquad (11.17)$$

for all $a,b \in J$ and $0 \le t \le 1$. This inequality is equivalent to the positivity of the **second divided difference**

$$\begin{aligned}
f[a,b,c] &= \frac{f(a)}{(a-b)(a-c)} + \frac{f(b)}{(b-a)(b-c)} + \frac{f(c)}{(c-a)(c-b)} \\
&= \frac{1}{c-b}\left(\frac{f(c)-f(a)}{c-a} - \frac{f(b)-f(a)}{b-a}\right) \qquad (11.18)
\end{aligned}$$

for every different $a,b,c \in J$.

Definition (11.17) makes sense if J is a convex subset of a vector space and f is a real functional defined on it. For a convex functional f the **Jensen inequality**

$$f\left(\sum_i t_i a_i\right) \le \sum_i t_i f(a_i) \qquad (11.19)$$

holds whenever $a_i \in J$ and for real numbers $t_i \ge 0$ and $\sum_i t_i = 1$.

A functional f is **concave** if $-f$ is convex.

Example 11.9. Using Theorem 11.9 and Example 11.7 we can prove

$$\operatorname{Tr} f(tA+(1-t)B) \le t\operatorname{Tr} f(A)+(1-t)\operatorname{Tr} f(B). \qquad (11.20)$$

for a convex function $f : \mathbb{R} \to \mathbb{R}$.

This property follows from the convexity of the function $t \mapsto \operatorname{Tr} f(C+tD)$. The first derivative is $\operatorname{Tr} f'(C+tD)D$ and the second one is

$$\operatorname{Tr} D\frac{\partial}{\partial t}f'(C+tD) = \sum_{ij} D_{ij}a_{ji}D_{ji}$$

if C is assumed to be diagonal, this is not a restriction, and $[a_{ij}]_{i,j=1}^n$ is the divided difference matrix of f'. From the convexity of f, $a_{ij} \ge 0$ and the derivative

$$\sum_{ij} a_{ji}|D_{ji}|^2$$

must be positive.

Since the function $\eta(x) := -x\log x$ is concave on $\mathbb{R}^+$, we can conclude that the **von Neumann entropy** $S(\rho) := -\text{Tr}\rho\log\rho$ is a concave functional on the set of density matrices.   □

Theorem 11.10. *If f_k and g_k are functions $[\alpha,\beta] \to \mathbb{R}$ such that for some $c_k \in \mathbb{R}$*

$$\sum_k c_k f_k(x) g_k(y) \geq 0$$

for every $x,y \in [\alpha,\beta]$, then

$$\sum_k c_k \text{Tr}\, f_k(A) g_k(B) \geq 0$$

whenever A,B are self-adjoint elements in $M_n(\mathbb{C})$ with eigenvalues in $[\alpha,\beta]$.

Proof. Let $A = \sum \lambda_i P_i$ and $B = \sum \mu_j Q_j$ be the spectral decompositions. Then

$$\sum_k c_k \text{Tr}\, f_k(A) g_k(B) = \sum_k \sum_{i,j} c_k \text{Tr}\, P_i f_k(A) g_k(B) Q_j$$
$$= \sum_{i,j} \text{Tr}\, P_i Q_j \sum_k c_k f_k(\lambda_i) g_k(\mu_j) \geq 0$$

as per the hypothesis.   □

In particular, if f is convex then

$$f(x) - f(y) - (x-y)f'(y) \geq 0$$

and

$$\text{Tr}\, f(A) \geq \text{Tr}\, f(B) + \text{Tr}\,(A - B)f'(B). \tag{11.21}$$

Replacing f by $t\log t$, we can see that the **relative entropy** of two states is positive:

$$S(\rho_1\|\rho_2) = \text{Tr}\,\rho_1\log\rho_1 - \text{Tr}\,\rho_1\log\rho_2 \geq \text{Tr}\,(\rho_2 - \rho_1).$$

This fact is the original **Klein inequality** (whose extension is the previous theorem). From the same theorem we can obtain a stronger estimate. Since

$$\eta(x) - \eta(y) - (x-y)\eta'(y) \geq \tfrac{1}{2}(x-y)^2,$$

we have

$$S(\rho_1\|\rho_2) \geq \tfrac{1}{2}\text{Tr}\,(\rho_1 - \rho_2)^2. \tag{11.22}$$

Example 11.10. From the inequality $1 + \log x \leq x \quad (x > 0)$ one obtains

$$\alpha^{-1}(a - a^{1-\alpha}b^\alpha) \leq a(\log a - \log b) \leq \alpha^{-1}(a^{1+\alpha}b^{-\alpha} - a)$$

for the numbers $a, b, \alpha > 0$. If A and B are positive invertible matrices, then Theorem 11.10 gives

$$\alpha^{-1}\mathrm{Tr}\,(A - A^{1-\alpha}B^\alpha) \leq \mathrm{Tr}\,A(\log A - \log B)$$
$$\leq \alpha^{-1}\mathrm{Tr}\,(A^{1+\alpha}B^{-\alpha} - A)$$

which provides a lower as well as an upper estimate for the relative entropy:

$$(1 - \alpha)S_\alpha(\rho_1\|\rho_2) \leq S(\rho_1\|\rho_2) \leq (1 - \alpha)S_{-\alpha}(\rho_1\|\rho_2). \tag{11.23}$$

11.4 Distances

The linear space $M_n(\mathbb{C})$ of complex $n \times n$ matrices is isomorphic to the space $\mathbb{C}^{n^2}$ and therefore

$$\left(\sum_{i,j}|A_{ij} - B_{ij}|^2\right)^{1/2}$$

is a natural distance between matrices $A, B \in M_n(\mathbb{C})$. This distance comes from the 2-norm which is a particular case of the p-norm defined as

$$\|X\|_p := \left(\mathrm{Tr}\,(X^*X)^{p/2}\right)^{1/p} \qquad (1 \leq p, X \in M_n(\mathbb{C})). \tag{11.24}$$

It was **von Neumann** who showed first that the Hölder inequality remains true in the matrix setting:

$$\|XY\|_1 \leq \|X\|_p\|Y\|_q \tag{11.25}$$

when $1/p + 1/q = 1$. An application of this inequality to **partial trace** is due to Carlen and Lieb.

Theorem 11.11. *Let $A \in B(\mathscr{H}_1 \otimes \mathscr{H}_2)$ be a positive operator. Then for all numbers $p \geq 1$, the inequality*

$$(\mathrm{Tr}_2(\mathrm{Tr}_1 A)^p)^{1/p} \leq \mathrm{Tr}_1((\mathrm{Tr}_2 A^p)^{1/p})$$

holds.

Proof. According to the Hölder inequality (11.25), we have

$$\mathrm{Tr}\,(B\mathrm{Tr}_1 A) \leq (\mathrm{Tr}\,B^q)^{1/q}(\mathrm{Tr}\,(\mathrm{Tr}_1 A)^p)^{1/p}$$

for positive $B \in B(\mathscr{H}_2)$, when $1/p + 1/q = 1$. Choose B such that equality holds and $\mathrm{Tr}\,B^q = 1$. Then

$$(\mathrm{Tr}\,(\mathrm{Tr}_1 A^p))^{1/p} = \mathrm{Tr}\,B\mathrm{Tr}_1(A) = \mathrm{Tr}\,(I \otimes B)A$$
$$= \sum_{i,j}\langle e_i \otimes f_j, (I \otimes B)A(e_i \otimes f_j)\rangle = \sum_{i,j}\langle e_i \otimes Bf_j, A(e_i \otimes f_j)\rangle$$

for any pair of orthonormal bases (e_i) and (f_j). We may assume that $B = \sum_j \lambda_j|f_j\rangle\langle f_j|$. The right-hand side can be estimated as follows.

$$\sum_{i,j}\lambda_i\langle e_i \otimes Bf_j, A(e_i \otimes f_j)\rangle \le \left(\sum_i \lambda_i^q\right)^{1/q}\sum_i\left(\sum_j\left(\langle e_i \otimes f_j, A(e_i \otimes f_j)\rangle\right)^p\right)^{1/p}$$
$$= \sum_i\left(\sum_j\left(\langle e_i \otimes f_j, A(e_i \otimes f_j)\rangle\right)^p\right)^{1/p}.$$

Since

$$\langle e_i \otimes f_j, A(e_i \otimes f_j)\rangle \le (\langle e_i \otimes f_j, A^p(e_i \otimes f_j)\rangle)^{1/p},$$

we obtain

$$(\mathrm{Tr}\,(\mathrm{Tr}_1 A^p))^{1/p} \le \sum_i\left(\sum_j\left(\langle e_i \otimes f_j, A^p(e_i \otimes f_j)\rangle\right)\right)^{1/p}$$
$$= \sum_i(\langle e_i, \mathrm{Tr}_2 A^p e_i\rangle)^{1/p}.$$

If the basis (e_i) consists of eigenvectors of $\mathrm{Tr}_2 A^p$, then the right-hand side is $\mathrm{Tr}\,(\mathrm{Tr}_2 A^p)^{1/p}$ and the proof is completed.   □

Let A be a positive operator in $B(\mathcal{H}_1 \otimes \mathcal{H}_2 \otimes \mathcal{H}_3)$. Then the inequality

$$\mathrm{Tr}_{1,3}(\mathrm{Tr}_2 A^r)^{1/r} \le \mathrm{Tr}_3(\mathrm{Tr}_2(\mathrm{Tr}_1 A)^r)^{1/r} \qquad (11.26)$$

can be deduced from the previous theorem for $0 < r \le 1$.

For $A \in M_n(\mathbb{C})$, the absolut value $|A|$ is defined as $\sqrt{A^*A}$ and it is a positive matrix. If A is self-adjoint and written in the form

$$A = \sum_i \lambda_i|e_i\rangle\langle e_i|,$$

where the vectors e_i form an orthonormal basis, then it is defined that

$$A_+ = \sum_{i:\lambda_i \ge 0}\lambda_i|e_i\rangle\langle e_i| \qquad A_- = -\sum_{i:\lambda_i \le 0}\lambda_i|e_i\rangle\langle e_i|. \qquad (11.27)$$

Then $A = A_+ - A_-$ and $|A| = A_+ + A_-$. The decomposition $A = A_+ - A_-$ is called the **Jordan decomposition** of A.

Lemma 11.1. *Let $\lambda_1 \ge \lambda_2 \ge \ldots \ge \lambda_n$ and $\mu_1 \ge \mu_2 \ge \ldots \ge \mu_n$ be the eigenvalues of the self-adjoint matrices A and B, respectively. Then*

$$\text{Tr}\,|A - B| \geq \sum_{i=1}^{n} |\lambda_i - \mu_i|.$$

Proof. Let $x - y$ be the Jordan decomposition of $A - B$. So $\text{Tr}\,|A - B| = \text{Tr}\,x + \text{Tr}\,y$. Consider $C = B + x = A + y$ with eigenvalues $v_1 \geq v_2 \geq \ldots \geq v_n$. Since $C \geq A, B$ we have

$$v_i \geq \lambda_i, \mu_i \qquad (i = 1, 2, \ldots, n)$$

which implies

$$2v_i - \lambda_i - \mu_i \geq |\lambda_i - v_i|.$$

Summing up, we obtain

$$\sum_{i=1}^{n} |\lambda_i - \mu_i| \leq \sum_{i=1}^{n} (2v_i - \lambda_i - \mu_i) = \text{Tr}\,(2C - A - B) = \text{Tr}\,x + \text{Tr}\,y$$

and the lemma is proven. $\qquad\qquad\qquad\qquad\qquad\qquad\qquad\qquad\qquad\qquad\qquad\square$

11.5 Majorization

For an n-tuple $t = (t_1, t_2, \ldots, t_n)$ of real numbers, $t^{\downarrow} := (t_1^{\downarrow}, t_2^{\downarrow}, \ldots, t_n^{\downarrow})$ denotes the decreasing rearrangement. Given $t, u \in \mathbb{R}^n$, we can say that t is **weakly majorized by** u if

$$\sum_{j=1}^{k} t_j^{\downarrow} \leq \sum_{j=1}^{k} u_j^{\downarrow} \qquad\qquad\qquad (11.28)$$

for every $1 \leq k \leq n$. In notation we can write $t \prec_w u$. If in addition for $j = n$ the equality holds, then t is said to be **majorized by** u, in notation $t \prec u$.

There are several equivalent characterizations of the majorization, without using rearrangement.

An $n \times n$ matrix $A = (a_{ij})$ is called **doubly stochastic** if $a_{ij} \geq 0$, $\sum_i a_{ij} = 1$ and $\sum_j a_{ij} = 1$, that is, the sum of each row and column is 1.

Example 11.11. Let $U = (u_{ij})$ be an $n \times n$ unitary matrix. Then $A = (|u_{ij}|^2)$ is doubly stochastic.

According to **Birkhoff's theorem** the extreme points of the set of all doubly stochastic matrices are the permutation matrices.

Theorem 11.12. *For $t, u \in \mathbb{R}^n$, the following statements are equivalent.*

(1) $\sum_{j=1}^{n} |t_j - x| \leq \sum_{j=1}^{n} |u_j - x|$ *for every $x \in \mathbb{R}$.*

(2) $t \prec u$.

(3) t belongs to the convex hull of the vectors obtained by permuting the coordinates of u.

(4) There exists a doubly stochastic matrix A such that t = Au.

Example 11.12. Let

$$\rho = \sum_{i=1}^{k} \lambda_i |z_i\rangle\langle z_i| = \sum_{i=1}^{k} \mu_i |w_j\rangle\langle w_i|$$

be decompositions of a density matrix ρ such that $|z_i\rangle$ and $|w_i\rangle$ are unit vectors. Assume that the first decomposition is orthogonal, that is, the Schmidt decomposition. Consider the unitary matrix constructed in the proof of Lemma 2.2:

$$\sum_{j=1}^{k} U_{ij}\sqrt{\lambda_j}|z_j\rangle = \sqrt{\mu_i}|w_i\rangle.$$

It is remarkable that

$$\mu_i = \sum_{j=1}^{k} |U_{ij}|^2 \lambda_j, \tag{11.29}$$

holds, that is, the probability vector $(\mu_i)_i$ is the image of the probability vector $(\lambda_j)_j$ under the bistochastic matrix $(|U_{ij}|^2)_{ij}$. Consequently,

$$(\mu_1, \mu_2, \dots, \mu_k) \prec (\lambda_1, \lambda_2, \dots, \lambda_k). \tag{11.30}$$

The k-tuple of the coefficients of an arbitrary decomposition is majorized by the k-tuple of the eigenvalues of the density matrix. This is a characteristic property of the Schmidt decomposition.   □

The n-tuples of real numbers may be regarded as diagonal matrices and the majorization can be extended to self-adjoint matrices. Suppose that $A, B \in M_n$ are so. Then $A \prec B$ means that the n-tuple of eigenvalues of A is majorized by the n-tuple of eigenvalues of B; similarly for the weak majorization.

The following result, known as **Ky Fan's maximum principle**, is sometimes useful to establish a (weak) majorization relation.

Theorem 11.13. *Let A be a self-adjoint matrix with decreasingly ordered eigenvalues $\lambda_1, \lambda_2, \dots, \lambda_n$. Then*

$$\sigma_k(A) := \sum_{i=1}^{k} \lambda_i = \sup\{\mathrm{Tr}\,AP : P = P^2 = P^*, \quad \mathrm{Tr}\,P = k\}$$
$$= \sup\{\mathrm{Tr}\,AC : 0 \le C \le I, \quad \mathrm{Tr}\,C \le k\}.$$

Since the majorization depends only on the spectrums, $A \prec B$ holds if and only if $UAU^* \prec VBV^*$ for some unitaries U and V. Therefore, it follows from Birkhoff's theorem that $A \prec B$ implies that

$$A = \sum_{i=1}^{n} p_i U_i B U_i^*$$ (11.31)

for some $p_i > 0$ with $\sum_i p_i = 1$ and for some unitaries U_i.

In the notation used in the previous theorem, the weak majorization $A \prec_w B$ is equivalent to $\sigma_k(A) \le \sigma_k(B)$ for all k. The majorization $A \prec B$ holds if and only if $A + aI \prec B + aI$ for some $a \in \mathbb{R}$. By shifting a self-adjoint matrix, we can make it to be positive always. When discussing the properties of majorization, we can restrict ourselves to positive (definite) matrices.

A positive unital mapping $\alpha : B(\mathcal{H}) \to B(\mathcal{H})$ is called **doubly stochastic** if $\mathrm{Tr} \circ \alpha = \mathrm{Tr}$. A state ρ_1 is called **more mixed** than the state ρ_2 if there exists a doubly stochastic map β such that $\rho_1 = \beta \rho_2$.

The following theorem tells that the more mixed relation of quantum states is essentially the same as the majorization relation of self-adjoint matrices. (Therefore, the majorization is mostly used.)

Theorem 11.14. *Let ρ_1 and ρ_2 be states. Then the following statements are equivalent.*

(1) $\rho_1 \prec \rho_2$.

(2) ρ_1 is more mixed than ρ_2.

(3) $\rho_1 = \sum_{i=1}^{n} \lambda_i U_i \rho_2 U_i^*$ for some convex combination λ_i and for some unitaries U_i.

(4) $\mathrm{Tr} f(\rho_1) \le \mathrm{Tr} f(\rho_2)$ for any convex function $f : \mathbb{R} \to \mathbb{R}$.

Since the function $\eta(t) = -t \log t$ is concave we can conclude that the more mixed state has larger von Neumann entropy. An example of doubly stochastic mapping is

$$[\beta(A)]_{ij} = \delta_{ij} A_{ij}$$ (11.32)

which tranforms matrices into diagonal ones. β is obviously positivity and trace preserving, hence sends density matrix into density matrix. The von Neummann entropy increases under this transformation.

Example 11.13. Let

$$\rho = \sum_{i=1}^{k} \lambda_i |z_i\rangle \langle z_i| = \sum_{i=1}^{k} \mu_i |w_j\rangle \langle w_i|$$

be decompositions of a density matrix ρ such that $|z_i\rangle$ and $|w_i\rangle$ are unit vectors. Assume that the first decomposition is orthogonal, that is, the Schmidt decomposition. Then

$$S(\rho) \equiv -\sum_{i=1}^{k} \lambda_i \log(\lambda_i) \le -\sum_{i=1}^{k} \mu_i \log(\mu_i)$$ (11.33)

is the consequence of Example 11.12. This relation was recognized by **E. T. Jaynes** in 1956 before the majorization appeared in quantum mechanics.   □

Theorem 11.15. *Let $f : I \to \mathbb{R}$ be a convex function defined on an interval I. If A and B are self-adjoint matrices with spectrum in I, then*

$$f(tA + (1-t)B) \prec_w t f(A) + (1-t)f(B)$$

for $0 < t < 1$.

This result extends (11.20).

Theorem 11.16. (Wehrl) *Let ρ be a density matrix of a finite quantum system $B(\mathcal{H})$ and $f : \mathbb{R}^+ \to \mathbb{R}^+$ a convex function with $f(0) = 0$. Then ρ is majorized by the density*

$$\rho_f = \frac{f(\rho)}{\operatorname{Tr} f(\rho)}.$$

Proof. Set $\lambda_1, \lambda_2, \ldots, \lambda_n$ for the decreasingly ordered eigenvalue list of ρ. Under the hypothesis on f, the inequality $f(x)y \leq f(y)x$ holds for $0 \leq x \leq y$. Hence for $i \leq j$, we have $\lambda_j f(\lambda_i) \geq \lambda_i f(\lambda_j)$ and

$$(f(\lambda_1) + \cdots + f(\lambda_k))(\lambda_{k+1} + \cdots + \lambda_n)$$
$$\geq (\lambda_1 + \cdots + \lambda_k)(f(\lambda_{k+1}) + \cdots + f(\lambda_n)).$$

Adding to both sides the term $(f(\lambda_1) + \cdots + f(\lambda_k))(\lambda_1 + \cdots + \lambda_k)$ we arrive at

$$(f(\lambda_1) + \cdots + f(\lambda_k)) \sum_{i=1}^n \lambda_i \geq (\lambda_1 + \cdots + \lambda_k) \sum_{i=1}^n f(\lambda_i).$$

This shows that the sum of the k largest eigenvalues of $f(\rho)/\operatorname{Tr} f(\rho)$ must exceed that of D, $\lambda_1 + \cdots + \lambda_k$. $\square$

Example 11.14. The canonical (Gibbs) state ρ_c^β at inverse temperature $\beta = (kT)^{-1}$ possesses the density $e^{-\beta H}/\operatorname{Tr} e^{-\beta H}$. Choosing $f(x) = x^{\beta'/\beta}$ with $\beta' > \beta$ the Theorem 11.16 tells us that

$$\rho_c^{\beta'} \succ \rho_c^\beta, \tag{11.34}$$

that is, at higher temperature the canonical state is more mixed. (The most mixed tracial state is canonical at infinite temperature.) According to the remark before the previous theorem, the entropy $S(\rho_c^\beta)$ is an increasing function of the temperature.
                                                                                                  $\square$

11.6 Operator Monotone Functions

Let $J \subset \mathbb{R}$ be an interval. A function $f : J \to \mathbb{R}$ is said to be **operator monotone** if $f(A) \leq f(B)$ whenever A and B are self-adjoint operators acting on the same Hilbert space, $A \leq B$ and their spectra are in J. One can see by an approximation argument that if the condition is fulfilled for matrices of arbitrarily large order, then it holds also for operators of infinite-dimensional Hilbert spaces. Note that an operator monotone function is automatically continuous and even smooth (see [20] about operator monotone functions).

Example 11.15. Let $t > 0$ be a parameter. The function $f(x) = -(t+x)^{-1}$ is operator monotone on $[0, \infty]$.

Let A and B positive invertible matrices of the same order. Then

$$A \le B \iff B^{-1/2}AB^{-1/2} \le I \iff \|B^{-1/2}AB^{-1/2}\| \le 1 \iff \|A^{1/2}B^{-1/2}\| \le 1.$$

Since the adjoint preserves the norm, the latest condition is equivalent to $\|B^{-1/2} A^{1/2}\| \le 1$, which implies that $B^{-1} \le A^{-1}$. □

Example 11.16. The function $f(x) = \log x$ is operator monotone on $(0, \infty)$.

This follows from the formula

$$\log x = \int_0^\infty \frac{1}{1+t} - \frac{1}{x+t} \, dt.$$

which is easy to verify. The integrand

$$f_t(x) := \frac{1}{1+t} - \frac{1}{x+t}$$

is operator monotone according to the previous example. The linear combination of operator monotone functions is operator monotone if the coefficients are positive. Taking the limit, we have a similar conclusion for an integral of operator monotone functions.

There are several other ways to show the operator monotonicity of the logarithm. □

Example 11.17. To show that the square root function is operator monotone, consider the function

$$F(t) := \sqrt{A + tX}$$

defined for $t \in [0, 1]$ and for fixed positive matrices A and X. If F is increasing, then $F(0) = \sqrt{A} \le \sqrt{A+X} = F(1)$.

In order to show that F is increasing, it is enough to see that the eigenvalues of $F'(t)$ are positive. Differentiating the equality $F(t)F(t) = A + tX$, we get

$$F'(t)F(t) + F(t)F'(t) = X.$$

As the limit of self-adjoint matrices, F' is self-adjoint and let $F'(t) = \sum_i \lambda_i E_i$ be its spectral decomposition. (Of course, both the eigenvalues and the projections depend on the value of t.) Then

$$\sum_i \lambda_i (E_i F(t) + F(t) E_i) = X$$

and after multiplication by E_j from the left and from the right, we have for the trace

$$2\lambda_j \operatorname{Tr} E_j F(t) E_j = \operatorname{Tr} E_j X E_j.$$

Since both traces are positive, λ_j must be positive as well.

The square root is matrix monotone for arbitrary matrix size and it follows by approximation that it is operator monotone as well.   □

The previous example contained an important idea. To decide about the operator monotonicity of a function f, one has to investigate the derivative of $f(A + tX)$.

Theorem 11.17. *A smooth function $f : J \to \mathbb{R}$ is operator monotone if and only if the divided difference matrix $D := [a_{ij}]_{i,j=1}^n$ defined as*

$$a_{ij} = \begin{cases} \dfrac{f(t_i) - f(t_j)}{t_i - t_j} & \text{if} \quad t_i - t_j \neq 0, \\[3mm] f'(t_i) & \text{if} \quad t_i - t_j = 0, \end{cases} \tag{11.35}$$

is positive semi-definite for all n and $t_1, t_2, \ldots, t_n \in J$.

It could be interesting to explain the background of this theorem. The argument given below is the essence of the proof. Let A be a self-adjoint and B be a positive semidefinite matrix. When f is operator monotone, the function $t \mapsto f(A + tB)$ is an increasing function of the real variable t. Therefore, the derivative, which is a matrix, must be positive semidefinite. To compute the derivative, we can use formula (11.9) (see Example 11.7).

It is easy to see that D must be positive if the Hadamard product $D \circ B$ is positive for every positive B. Therefore, the positivity of the derivative in all positive directions implies the positivity of the divided difference matrix. We can conclude that the latter is a necessary condition for the operator monotonicity of f.

To show that this condition is sufficient, we need to show that the derivative $D \circ B$ is positive semidefinite for all positive semidefinite B. This follows from the **Schur therem**, which claims that the Hadamard product of positive semidefinite matrices is positive semidefinite.

Example 11.18. The function $f(x) := \exp x$ is not operator monotone, since the divided difference matrix

$$\begin{pmatrix} \exp x & \dfrac{\exp x - \exp y}{x - y} \\[3mm] \dfrac{\exp y - \exp x}{y - x} & \exp y \end{pmatrix}$$

does not have a positive determinant.   □

Operator monotone functions on $\mathbb{R}^+$ have a special integral representation

$$f(x) = f(0) + \beta x + \int_0^\infty \frac{\lambda x}{\lambda + x} d\mu(\lambda), \tag{11.36}$$

where μ is a measure such that

$$\int_0^\infty \frac{\lambda}{\lambda + 1} d\mu(\lambda)$$

is finite. This result is called **Löwner theorem**. Since the integrand

$$\frac{\lambda x}{\lambda + x} = \lambda - \frac{\lambda^2}{\lambda + x}$$

is an operator monotone function of x (see Example 11.15), one part of the Löwner theorem is straightforward.

It is not always easy to decide if a function is operator monotone. An efficient method is based on holomorphic extension. The set $\mathbb{C}_+ := \{a + ib : a, b \in \mathbb{R} \text{ and } b > 0\}$ is called "upper half-plain." A function $J \to \mathbb{R}$ is operator monotone if and only if it has a holomorphic extension to the upper half-plain such that its range is in $\mathbb{C}_+$.

Example 11.19. The representation

$$x^t = \frac{\sin \pi t}{\pi} \int_0^\infty \frac{\lambda^{t-1} x}{\lambda + x} d\lambda \tag{11.37}$$

shows that $f(x) = x^t$ is operator monotone when $0 < t < 1$. In other words,

$$0 \le A \le B \quad \text{imply} \quad A^t \le B^t,$$

which is often called **Löwner–Heinz inequality**.

We can arrive at the same conclusion by holomorphic extension. If

$$a + ib = Re^{\varphi i} \quad \text{with} \quad 0 \le \varphi \le \pi,$$

then $a + ib \mapsto R^t e^{t\varphi i}$ is holomorphic and it maps $\mathbb{C}_+$ into itself when $0 \le t \le 1$. This shows that $f(x) = x^t$ is operator monotone for these values of the parameter but not for any other value. □

Example 11.20. Let

$$f_p(x) = p(1 - p) \frac{(x - 1)^2}{(x^p - 1)(x^{1-p} - 1)}. \tag{11.38}$$

This function is operator monotone if $-1 < p < 2$. The key idea of the proof is an integral representation:

$$\frac{1}{f_p(x)} = \frac{\sin p\pi}{\pi} \int_0^\infty d\lambda \, \lambda^{p-1} \int_0^1 ds \int_0^1 dt \frac{1}{x((1-t)\lambda + (1-s)) + (t\lambda + s)}, \tag{11.39}$$

if $0 < p < 1$, the other values of p can be treated similarly. The integrand is operator monotone decreasing (as a function of t, s and λ) and so is the triple integral. Since $f_p(x)^{-1}$ is operator monotone decreasing, $f_p(x)$ is operator monotone.   □

Theorem 11.18. *Let $f : \mathbb{R}^+ \to \mathbb{R}$ be an operator monotone function. For positive matrices K and L, let P be the projection onto the range of $(K-L)_+$. Then*

$$\operatorname{Tr} PL(f(K) - f(L)) \geq 0. \tag{11.40}$$

Proof. From the integral representation

$$f(x) = \int_0^\infty \frac{x(1+s)}{x+s} \, d\mu(s)$$

we have

$$\operatorname{Tr} PL(f(K) - f(L)) = \int_0^\infty (1+s)s \operatorname{Tr} PL(K+s)^{-1}(K-L)(L+s)^{-1} \, d\mu(s).$$

Hence it is sufficient to prove that

$$\operatorname{Tr} PL(K+s)^{-1}(K-L)(L+s)^{-1} \geq 0$$

for $s > 0$. Let $\Delta_0 := K - L$ and observe the integral representation

$$(K+s)^{-1}\Delta_0(L+s)^{-1} = \int_0^1 s(L+t\Delta_0+s)^{-1}\Delta_0(L+t\Delta_0+s)^{-1} \, dt.$$

So we can make another reduction:

$$\operatorname{Tr} PL(L+t\Delta_0+s)^{-1}t\Delta_0(L+t\Delta_0+s)^{-1} \geq 0$$

is enough to be shown. If $C := L+t\Delta_0$ and $\Delta := t\Delta_0$, then $L = C - \Delta$ and we have

$$\operatorname{Tr} P(C-\Delta)(C+s)^{-1}\Delta(C+s)^{-1} \geq 0. \tag{11.41}$$

Let us write our operators in the form of 2×2 block-matrices:

$$(C+s)^{-1} = \begin{bmatrix} V_1 & V_2 \\ V_2^* & V_3 \end{bmatrix}, \quad P = \begin{bmatrix} I & 0 \\ 0 & 0 \end{bmatrix}, \quad \Delta = \begin{bmatrix} \Delta_+ & 0 \\ 0 & -\Delta_- \end{bmatrix}.$$

The left-hand side of the inequality (11.41) can then be rewritten as

$$\begin{aligned}
\operatorname{Tr} P(C-\Delta)(V\Delta V) &= \operatorname{Tr}[(C-\Delta)(V\Delta V)]_{11} \\
&= \operatorname{Tr}[(V^{-1}-\Delta-s)(V\Delta V)]_{11} \\
&= \operatorname{Tr}[\Delta V - (\Delta+s)(V\Delta V)]_{11}
\end{aligned}$$

$$= \text{Tr}\,(\Delta_+ V_{11} - (\Delta_+ + s)(V\Delta V)_{11})$$
$$= \text{Tr}\,(\Delta_+ (V - V\Delta V)_{11} - s(V\Delta V)_{11}). \tag{11.42}$$

Because of the positivity of L, we have $V^{-1} \geq \Delta + s$, which implies $V = VV^{-1}V \geq V(\Delta + s)V = V\Delta V + sV^2$. As the diagonal blocks of a positive operator are themselves positive, this further implies

$$V_1 - (V\Delta V)_{11} \geq s(V^2)_{11}.$$

Inserting this in (11.42) gives

$$\begin{aligned}
\text{Tr}\,[(V^{-1} - \Delta - s)(V\Delta V)]_{11} &= \text{Tr}\,(\Delta_+ (V - V\Delta V)_{11} - s(V\Delta V)_{11}) \\
&\geq \text{Tr}\,(\Delta_+ s(V^2)_{11} - s(V\Delta V)_{11}) \\
&= s\text{Tr}\,(\Delta_+ (V^2)_{11} - (V\Delta V)_{11}) \\
&= s\text{Tr}\,(\Delta_+ (V_1 V_1 + V_2 V_2^*) - (V_1 \Delta_+ V_1 - V_2 \Delta_- V_2^*) \\
&= s\text{Tr}\,(\Delta_+ V_2 V_2^* + V_2 \Delta_- V_2^*))
\end{aligned}$$

This quantity is positive. $\qquad\qquad\square$

Theorem 11.19. *Let A and B be positive operators, then for all $0 \leq s \leq 1$,*

$$2\text{Tr}\,A^s B^{1-s} \geq \text{Tr}\,(A + B - |A - B|). \tag{11.43}$$

Proof. In the proof we may assume that $0 \leq s \leq 1/2$ and apply the previous theorem to the case $f(x) = x^{s/(1-s)}$, $K = A^{1-s}$ and $L = B^{1-s}$, since $0 \leq s/(1-s) \leq 1$ and $f(x)$ is operator monotone. With P the projection on the range of $(A^{1-s} - B^{1-s})_+$, this yields

$$\text{Tr}\,PB^{1-s}(A^s - B^s) = \text{Tr}\,PB^{1-s}A^s - \text{Tr}\,PB \geq 0.$$

A simple rearrangment yields

$$\text{Tr}\,A^s P(A^{1-s} - B^{1-s}) \leq \text{Tr}\,P(A - B). \tag{11.44}$$

Since P is the projection on the range of the positive part of $(A^{1-s} - B^{1-s})$, the left-hand side can be rewritten as $\text{Tr}\,A^s(A^{1-s} - B^{1-s})_+$ and

$$\text{Tr}\,A^s(A^{1-s} - B^{1-s}) \leq \text{Tr}\,A^s(A^{1-s} - B^{1-s})_+ = \text{Tr}\,A^s P(A^{1-s} - B^{1-s}).$$

On the other hand, the right-hand side of (11.44) is upper-bounded by $\text{Tr}\,(A - B)_+$; thus we have

$$\text{Tr}\,(A - A^s B^{1-s}) \leq \text{Tr}\,(A - B)_+ = \frac{1}{2}\text{Tr}\,((A - B) + |A - B|).$$

This is equivalent to the statement. $\qquad\qquad\square$

Operator monotone functions on $\mathbb{R}^+$ may be used to define **positive operator means**. A theory of means of positive operators was developed by Kubo and Ando [69]. Their theory has many interesting applications. Here I do not go into the details concerning operator means but let us confine ourselves to the essentials and restrict ourselves to matrices mostly. Operator means are binary operations on positive operators acting on a Hilbert space and they satisfy the following conditions.

(1) $M(A,A) = A$ for every A,

(2) $M(A,B) = M(B,A)$ for every A and B,

(3) if $A \leq B$, then $A \leq M(A,B) \leq B$,

(4) if $A \leq A'$ and $B \leq B'$, then $M(A,B) \leq M(A',B')$,

(5) M is continuous.

Note that the above conditions are not independent, (1) and (4) imply (3).

An important further requirement is the **transformer inequality**:

(6) $CM(A,B)C^* \leq M(CAC^*, CBC^*)$

for all not-necessary self-adjoint operators C.

The key issue of the theory is that operator means are in a 1-to-1 correspondence with operator monotone functions satisfying conditions $f(1) = 1$ and $tf(t^{-1}) = f(t)$. Given an operator monotone function f, the corresponding mean is

$$M_f(A,B) = A^{1/2}f(A^{-1/2}BA^{-1/2})A^{1/2} \tag{11.45}$$

when A is invertible. (When A is not invertible, take a sequence A_n of invertible operators approximating A and let $M_f(A,B) = \lim_n M_f(A_n,B)$.)

An important example is the **geometric mean**

$$A\#B = A^{1/2}(A^{-1/2}BA^{-1/2})^{1/2}A^{1/2} \tag{11.46}$$

which corresponds to $f(x) = \sqrt{x}$. The geometric mean $A\#B$ is the unique positive solution of the equation

$$XA^{-1}X = B \tag{11.47}$$

and therefore $(A\#B)^{-1} = A^{-1}\#B^{-1}$.

Theorem 11.20. *Let A and B be $n \times n$ positive matrices. The geometric mean $A\#B$ is the largest self-adjoint matrix X such that the block-matrix*

$$\begin{bmatrix} A & X \\ X & B \end{bmatrix}$$

is positive.

The relevance of the operator means to quantum information theory is demonstrated by the following example.

Example 11.21. Let ω and ρ be density matrices. Their **fidelity** can be written as

$$F(\omega,\rho) = \operatorname{Tr}\omega(\omega^{-1}\#\rho).\tag{11.48}$$

From this observation the symmetry of the fidelity follows.

$$F(\omega,\rho) = \operatorname{Tr}\omega(\omega^{-1}\#\rho) = \operatorname{Tr}(\omega^{-1}\#\rho)\omega(\omega^{-1}\#\rho)(\omega^{-1}\#\rho)^{-1}$$
$$= \operatorname{Tr}\rho(\omega^{-1}\#\rho)^{-1} = \operatorname{Tr}\rho(\omega\#\rho^{-1}) = F(\rho,\omega)$$

$\square$

Example 11.22. Let ω and ρ be density matrices. The function $f(x) = \log x$ is operator monotone but $tf(t^{-1}) = f(t)$ does not hold, so (11.45) gives an asymmetric mean.

$$\operatorname{Tr}M_f(\omega,\rho) = \operatorname{Tr}\omega\log(\omega^{-1/2}\rho\omega^{-1/2}).\tag{11.49}$$

This is a relative entropy quantity studied by Belavkin and Staszewski [13] (up to sign). Hiai and Petz proved the inequality

$$\operatorname{Tr}\omega\log(\omega^{-1/2}\rho\omega^{-1/2}) \geq \operatorname{Tr}\omega(\log\rho - \log\omega).\tag{11.50}$$

$\square$

Let $J \subset \mathbb{R}$ be an interval. A function $f : J \to \mathbb{R}$ is said to be **operator convex** if

$$f(tA + (1-t)B) \leq tf(A) + (1-t)f(B)\tag{11.51}$$

for all self-adjoint operators A and B whose spectra are in J and for all numbers $0 \leq t \leq 1$. f is operator concave if $-f$ is operator convex.

Since self-adjoint operators may be approximated by self-adjoint matrices, (11.51) holds for operators when it holds for matrices. It is a standard argument that the convex combination $(tA + (1-t)B)$ can be replaced by a more general one, $\sum_{k=1}^{n} t_k A_k$, where $\sum_{k=1}^{n} t_k = 1$ and $t_k \geq 0$. However, it is not at all trivial that one may consider operator coefficients as well.

Theorem 11.21. *Let f be an operator-convex function and let $T_1, T_2, \ldots, T_n$ be operators (acting on the same space) such that*

$$\sum_{k=1}^{n} T_k^* T_k = I.$$

Then

$$f\left(\sum_{i=k}^{n} T_k^* A_k T_k\right) \leq \sum_{k=1}^{n} T_k^* f(A_k) T_k.\tag{11.52}$$

Since the sum $\sum_{k=1}^{n} T_k^* A_k T_k$ is sometimes called **C*-convex combination**, one can say that an operator-convex function is C*-convex as well.

Proof. of the theorem: To show the inequality T_k's are regared as a column of a unitary matrix (with operator entries). There exists an $n \times n$ unitary matrix $(U_{ij})_{i,j=1}^{n}$ such that (T_i) is its last column, that is, $T_k = U_{kn}$. Then

$$\sum_{i=k}^{n} T_k^* A_k T_k = \left(U^* \mathrm{Diag}\,(A_1, A_2, \ldots, A_n) U \right)_{nn}.$$

Let

$$W = \mathrm{Diag}\,(\theta, \theta^2, \ldots, \theta^{n-1}, 1),$$

where $\theta = \exp(2\pi i / n)$. We have

$$f\left(\sum_{i=k}^{n} T_k^* A_k T_k \right) = f\left(\left(\sum_{k=1}^{n} \frac{1}{n} W^{-k} (U^* \mathrm{Diag}\,(A_1, A_2, \ldots, A_n) U) W^k \right)_{nn} \right)$$

$$= f\left(\sum_{k=1}^{n} \frac{1}{n} W^{-k} (U^* \mathrm{Diag}\,(A_1, A_2, \ldots, A_n) U) W^k \right)_{nn}$$

$$\leq \left(\sum_{k=1}^{n} \frac{1}{n} f(W^{-k} U^* \mathrm{Diag}\,(A_1, A_2, \ldots, A_n) U W^k) \right)_{nn}$$

$$= \left(\sum_{k=1}^{n} \frac{1}{n} W^{-k} U^* \mathrm{Diag}\,(f(A_1), f(A_2), \ldots, f(A_n)) U W^k \right)_{nn}$$

$$= \left(U^* \mathrm{Diag}\,(f(A_1), f(A_2), \ldots, f(A_n)) U \right)_{nn} = \sum_{k=1}^{n} T_k^* f(A_k) T_k.$$

and conclude (11.52).   □

Example 11.23. The function $f(t) = t^2$ is operator convex on the whole real line. This follows from the obvious inequality

$$\left(\frac{A+B}{2} \right)^2 \leq \frac{A^2 + B^2}{2}.$$

It is remarkable that

$$(T_1^* A T_1 + T_2^* B T_2)^2 \leq T_1^* A^2 T_1 + T_2^* B^2 T_2$$

does not seem to have an easy direct proof. (This inequality is a particular case of (11.52).)   □

Corollary 11.1. *Let f be an operator-convex function on an interval J such that $0 \in J$. If $\|V\| \leq 1$ and $f(0) \leq 0$, then*

$$f(V^* A V) \leq V^* f(A) V \tag{11.53}$$

if the spectrum of $A = A^*$ lies in J.

Proof. Choose $B = 0$ and W such that $V^*V + W^*W = I$. Then

$$f(V^*AV + W^*BW) \leq V^*f(A)V + W^*f(B)W$$

holds and gives our statement.   □

Example 11.24. The function $f(x) = (x+t)^{-1}$ is operator convex on $[0, \infty]$ when $x > 0$. It is enough to show that

$$\left(\frac{A+B}{2}\right)^{-1} \geq \frac{A^{-1} + B^{-1}}{2}$$

which is equivalent with

$$\left(\frac{B^{-1/2}AB^{-1/2} + I}{2}\right)^{-1} \geq \frac{(B^{-1/2}AB^{-1/2})^{-1} + I}{2}.$$

This holds, since

$$\left(\frac{X+I}{2}\right)^{-1} \geq \frac{X^{-1} + I}{2}$$

is true for an invertible operator $X \geq 0$.   □

The integral formula in Example 11.16 gives that $f(x) = \log x$ is operator concave; however, much more is true. It follows from the Löwner theorem that any operator monotone function on $\mathbb{R}^+$ is operetor concave.

11.7 Positive Mappings

Let $\alpha : B(\mathcal{H}) \to B(\mathcal{K})$ be a linear mapping for finite-dimensional Hilbert spaces $\mathcal{H}$ and $\mathcal{K}$. α is called **positive** if it sends positive (semidefinite) operators to positive (semidefinite) operators.

Theorem 11.22. *Let $\alpha : M_n(\mathbb{C}) \to M_k(\mathbb{C})$ be a positive unital linear mapping and $f : \mathbb{R} \to \mathbb{R}$ be a convex function. Then*

$$\mathrm{Tr}\, f(\alpha(A)) \leq \mathrm{Tr}\, \alpha(f(A))$$

for every $A \in M_n(\mathbb{C})^{sa}$.

Proof. By means of the spectral decompositions $A = \sum_j v_j Q_j$ and $\alpha(A) = \sum_i \mu_i P_i$ we have

$$\mu_i = \mathrm{Tr}\,(\alpha(A)P_i)/\mathrm{Tr}\,P_i = \sum_j v_j \mathrm{Tr}\,(\alpha(Q_j)P_i)/\mathrm{Tr}\,P_i ,$$

whereas the convexity of f yields

$$f(\mu_i) \le \sum_j f(\nu_j) \mathrm{Tr}\left(\alpha(Q_j)P_i\right)/\mathrm{Tr}\,P_i\,.$$

Therefore,

$$\mathrm{Tr}\,f(\alpha(A)) = \sum_i f(\mu_i)\mathrm{Tr}\,P_i \le \sum_{i,j} f(\nu_j)\mathrm{Tr}\left(\alpha(Q_j)P_i\right) = \mathrm{Tr}\,\alpha(f(A))\,,$$

which was to be proven.   □

If we apply this result to the mapping $\alpha : M_n(\mathbb{C}) \oplus M_n(\mathbb{C}) \to M_n(\mathbb{C})$ defined by

$$\alpha(A \oplus B) = tA + (1-t)B$$

for some $0 < t < 1$, then we obtain the convexity

$$\mathrm{Tr}\,f(tA + (1-t)B) \le t\mathrm{Tr}\,f(A) + (1-t)\mathrm{Tr}\,f(B)\,. \tag{11.54}$$

Example 11.25. Let $\mathscr{E} : M_n(\mathbb{C}) \to M_n(\mathbb{C})$ be a positive unital trace-preserving mapping. Therefore $\mathscr{E}$ sends the density matrix ρ into a density matrix $\mathscr{E}(\rho)$. It can be shown that

$$S(\mathscr{E}(\rho)) \ge S(\rho)\,.$$

Our argument is based on the fact that $-S(\rho)$ can be obtained by differentiating $\mathrm{Tr}\,\rho^p$. If $p > 1$, then $f(t) = t^p$ is convex, hence

$$\mathrm{Tr}\left(\mathscr{E}(\rho)^p - \mathscr{E}(\rho)\right) \le \mathrm{Tr}(\rho^p - \rho)$$

according to Theorem 11.22. Dividing by $p-1$ and letting $p \to 1$, we obtain the statement.   □

The **dual** $\alpha^* : B(\mathscr{K}) \to B(\mathscr{H})$ of α is defined by the equation

$$\mathrm{Tr}\,\alpha(A)B = \mathrm{Tr}\,A\alpha^*(B) \qquad (A \in B(\mathscr{H}), B \in B(\mathscr{K}))\,. \tag{11.55}$$

It is easy to see that α is positive if and only if α^* is positive and α is trace preserving if and only if α^* is unit preserving.

The operator inequality

$$\alpha(AA^*) \ge \alpha(A)\alpha(A)^*$$

is called **Schwarz inequality**. If α is a positive mapping, then the ineqiality holds for normal operators. This result is called **Kadison inequality**. (The Schwarz inequality holds for an arbitrary operator A if α is completely positive, or 2-positive.)

Theorem 11.23. *Let $\alpha : B(\mathscr{H}) \to B(\mathscr{K})$ be a positive unit-preserving mapping.*

(1) If $A \in B(\mathcal{H})$ is a normal operator (that is, $A^*A = AA^*$), then

$$\alpha(AA^*) \geq \alpha(A)\alpha(A)^*.$$

(2) If $A \in B(\mathcal{H})$ is a positive operator such that A and $\alpha(A)$ are invertible, then

$$\alpha(A^{-1}) \geq \alpha(A)^{-1}.$$

Proof. A has a spectral decomposition $\sum_i \lambda_i P_i$, where P_i's are pairwise orthogonal projections. We have $A^*A = \sum_i |\lambda_i|^2 P_i$ and

$$\begin{bmatrix} I & \alpha(A) \\ \alpha(A)^* & \alpha(A^*A) \end{bmatrix} = \sum_i \begin{bmatrix} 1 & \lambda_i \\ \overline{\lambda_i} & |\lambda_i|^2 \end{bmatrix} \otimes \alpha(P_i)$$

Since $\alpha(P_i)$ is positive, the left-hand side is positive as well. Reference to Theorem 11.6 gives the first inequality.

To prove the second inequality, use the identity

$$\begin{bmatrix} \alpha(A) & I \\ I & \alpha(A^{-1}) \end{bmatrix} = \sum_i \begin{bmatrix} \lambda_i & 1 \\ 1 & \lambda_i^{-1} \end{bmatrix} \otimes \alpha(P_i)$$

to conclude that the left-hand side is a positive block-matrix. The positivity implies our statement. □

If we fix bases in $\mathcal{H}$ and in $\mathcal{K}$, then $B(\mathcal{H})$ can be identified by a matrix algebra $M_n(\mathbb{C})$, and similarly $B(\mathcal{K})$ is $M_k(\mathbb{C})$ (when n is the dimension of $\mathcal{H}$ and k is the dimension of $\mathcal{K}$).

Theorem 11.24. *Let $\mathscr{E} : M_n(\mathbb{C}) \to M_k(\mathbb{C})$ be a linear mapping. Then the following conditions are equivalent.*

(1) $\mathscr{E} \otimes \mathrm{id}_n$ is a positive mapping, when $\mathrm{id}_n : M_n(\mathbb{C}) \to M_n(\mathbb{C})$ is the identity mapping.
(2) The block-matrix X defined by

$$X_{ij} = \mathscr{E}(E_{ij}) \qquad (1 \leq i, j \leq n) \tag{11.56}$$

is positive.
(3) There are operators $V_t : \mathbb{C}^n \to \mathbb{C}^k$ $(1 \leq t \leq k^2)$ such that

$$\mathscr{E}(A) = \sum_t V_t A V_t^*. \tag{11.57}$$

(4) For finite families $A_i \in M_n(\mathbb{C})$ and $B_i \in M_k(\mathbb{C})$ $(1 \leq i \leq n)$, the inequality

$$\sum_{i,j} B_i^* \mathscr{E}(A_i^* A_j) B_j \geq 0$$

holds.

Proof. (1) implies (2): The matrix

$$\sum_{i,j} E_{ij} \otimes E_{ij} = \frac{1}{n}\left(\sum_{i,j} E_{ij} \otimes E_{ij}\right)^2$$

is positive. Therefore,

$$\left(\mathrm{id}_n \otimes \mathcal{E}\right)\left(\sum_{i,j} E_{ij} \otimes E_{ij}\right) = \sum_{i,j} E_{ij} \otimes \mathcal{E}(E_{ij}) = X$$

is positive as well.

(2) implies (3): Assume that the block-matrix X is positive. There are orthogonal projections P_i $(1 \le i \le n)$ on $\mathbb{C}^{nk}$ such that they are pairwise orthogonal and

$$P_i X P_j = \mathcal{E}(E_{ij}).$$

We have a decomposition

$$X = \sum_{t=1}^{nk} |f_t\rangle\langle f_t|,$$

where $|f_t\rangle$ are appropriately normalized eigenvectors of X. Since P_i is a partition of unity, we have

$$|f_t\rangle = \sum_{i=1}^{n} P_i |f_t\rangle$$

and set $V_t : \mathbb{C}^n \to \mathbb{C}^k$ by

$$V_t|s\rangle = P_s|f_t\rangle.$$

($|s\rangle$ are the canonical basis vectors.) In this notation

$$X = \sum_{t}\sum_{i,j} P_i|f_t\rangle\langle f_t|P_j = \sum_{i,j} P_i\left(\sum_{t} V_t|i\rangle\langle j|V_t^*\right) P_j$$

and

$$\mathcal{E}(E_{ij}) = P_i X P_j = \sum_{t} V_t E_{ij} V_t^*.$$

Since this holds for all matrix units E_{ij}, we obtain

$$\mathcal{E}(A) = \sum_{t} V_t A V_t^*.$$

(3) implies (4): Assume that $\mathcal{E}$ is in the form of (11.57). Then

$$\sum_{i,j} B_i^* \mathscr{E}(A_i^* A_j) B_j = \sum_t \sum_{i,j} B_i^* V_t(A_i^* A_j) V_t^* B_j$$

$$= \sum_t \left(\sum_i A_i V_t^* B_i \right)^* \left(\sum_j A_j V_t^* B_j \right) \geq 0$$

follows.

(4) implies (1): We have

$$\mathscr{E} \otimes \mathrm{id}_n : M_n(B(\mathscr{H})) \to M_n(B(\mathscr{K})).$$

Since any positive operator in $M_n(B(\mathscr{H}))$ is the sum of operators in the form $\sum_{i,j} A_i^* A_j \otimes E_{ij}$ (Theorem 11.8), it is enough to show that

$$X := \mathscr{E} \otimes id_n \left(\sum_{i,j} A_i^* A_j \otimes E_{ij} \right) = \sum_{i,j} \mathscr{E}(A_i^* A_j) \otimes E_{ij}$$

is positive. On the other hand, $X \in M_n(B(\mathscr{K}))$ is positive if and only if

$$\sum_{i,j} B_i^* X_{ij} B_j = \sum_{i,j} B_i^* \mathscr{E}(A_i^* A_j) B_j \geq 0.$$

The positivity of this operator is supposed in (4), hence (1) is shown. □

When the linear mapping $\mathscr{E}$ is between abstract C^*-algebras, then the equivalent conditions (1) and (4) can be used.

Example 11.26. A typical completely positive mapping is the **partial trace**. Let $\mathscr{H}$ and $\mathscr{K}$ be Hilbert spaces and (f_i) be a basis in $\mathscr{K}$. For each i set a linear operator $V_i : \mathscr{H} \to \mathscr{H} \otimes \mathscr{K}$ as $V_i e = e \otimes f_i$ ($e \in \mathscr{H}$). These operators are isometries with pairwise orthogonal ranges and the adjoints act as $V_i^*(e \otimes f) = \langle f_i, f \rangle e$. The linear mapping

$$\mathrm{Tr}_2 : B(\mathscr{H} \otimes \mathscr{K}) \to B(\mathscr{H}), \qquad A \mapsto \sum_i V_i A V_i^* \qquad (11.58)$$

is called **partial trace** over the second factor. The reason for that is the formula

$$\mathrm{Tr}_2(X \otimes Y) = X \mathrm{Tr} Y. \qquad (11.59)$$

Note that the partial trace is a conditional expectation up to a constant factor; its complete positivity follows also from this fact (see (9.6)). □

Example 11.27. The trace $\mathrm{Tr} : M_k(\mathbb{C}) \to \mathbb{C}$ is completely positive if $\mathrm{Tr} \otimes \mathrm{id}_n : M_k(\mathbb{C}) \otimes M_n(\mathbb{C}) \to M_n(\mathbb{C})$ is a positive mapping. However, this is a partial trace which is known to be positive (even completely positive).

It follows that any positive linear functional $\psi : M_k(\mathbb{C}) \to \mathbb{C}$ is completely positive. Since $\psi(A) = \mathrm{Tr} \rho A$ with a certain positive ρ, ψ is the composition of the completely positive mappings $A \mapsto \rho^{1/2} A \rho^{1/2}$ and Tr. □

Example 11.28. Let $\mathscr{E} : M_n(\mathbb{C}) \to M_k(\mathbb{C})$ be a positive linear mapping such that $\mathscr{E}(A)$ and $\mathscr{E}(B)$ commute for any $A, B \in M_n(\mathbb{C})$. We can show that $\mathscr{E}$ is completely positive as follows.

Any two self-adjoint matrices in the range of $\mathscr{E}$ commute, so we can change the basis such that all of them become diagonal. It follows that $\mathscr{E}$ has the form

$$\mathscr{E}(A) = \sum_i \psi_i(A) E_{ii},$$

where E_{ii} are the diagonal matrix units and ψ_i are positive linear functionals. Since the sum of completely positive mappings is completely positive, it is enough to show that $A \mapsto \psi(A)F$ is completely positive for a positive functional ψ and for a positive matrix F. The complete positivity of this mapping means that for an $m \times m$ block-matrix X with entries $X_{ij} \in M_n(\mathbb{C})$, the block-matrix $(\psi(X_{ij})F)_{i,j=1}^n$ should be positive. This is true, since the matrix $(\psi(X_{ij}))_{i,j=1}^n$ is positive (due to the complete positivity of ψ). $\square$

The next result tells that the **Kraus representation** of a completely positive mapping is unique up to a unitary matrix.

Theorem 11.25. *Let $\mathscr{E} : M_n(\mathbb{C}) \to M_m(\mathbb{C})$ be a linear mapping which is represented as*

$$\mathscr{E}(A) = \sum_{t=1}^k V_t A V_t^* \quad and \quad \mathscr{E}(A) = \sum_{t=1}^k W_t A W_t^*.$$

Then there exist a $k \times k$ unitary matrix (c_{tu}) such that

$$W_t = \sum_u c_{tu} V_u.$$

Proof. Let x_i be a basis in $\mathbb{C}^m$ and y_j be a basis in $\mathbb{C}^n$. Consider the vectors

$$v_t := \sum_{ij} x_i \otimes V_t y_j \quad and \quad w_t := \sum_{ij} x_i \otimes W_t y_j.$$

We have

$$|v_t\rangle\langle v_t| = \sum_{iji'j'} |x_i\rangle\langle x_{i'}| \otimes V_t |y_j\rangle\langle y_{j'}| V_t^*$$

and

$$|w_t\rangle\langle w_t| = \sum_{iji'j'} |x_i\rangle\langle x_{i'}| \otimes W_t |y_j\rangle\langle y_{j'}| W_t^*.$$

Our hypothesis implies that

$$\sum_t |v_t\rangle\langle v_t| = \sum_t |w_t\rangle\langle w_t|.$$

Lemma 2.2 tells us that there is a unitary matrix (c_{tu}) such that

$$w_t = \sum_u c_{tu} v_u.$$

This implies that

$$\langle x_i | W_t | y_j \rangle = \langle x_i | \sum_u c_{tu} V_u | y_j \rangle$$

for every i and j and the statement of the theorem can be concluded.   □

11.8 Matrix Algebras

Let $\mathscr{H}$ be a Hilbert space. The set of bounded operators $B(\mathscr{H})$ has an algebraic structure, $B(\mathscr{H})$ is a unital *-algebra. If the space $\mathscr{H}$ is n-dimensional, then $\mathscr{B} := B(\mathscr{H})$ can be identified with the algebra $M_n(\mathbb{C})$ of $n \times n$ complex matrices. The subalgebras of $\mathscr{B}$ containing the identity and closed under adjoints will be called **matrix algebras**.

Example 11.29. Assume that $\mathscr{A}$ is a subalgebra of $\mathscr{B}$. If we choose the basis of $\mathscr{H}$ properly, then elements of $\mathscr{A}$ can be written in block-diagonal form, for example as

$$\begin{bmatrix} A & 0 & 0 & 0 & 0 \\ 0 & A & 0 & 0 & 0 \\ 0 & 0 & A & 0 & 0 \\ 0 & 0 & 0 & B & 0 \\ 0 & 0 & 0 & 0 & C \end{bmatrix} = (I_3 \otimes A) \oplus B \oplus C, \qquad (11.60)$$

where, say, A is an $r \times r$ matrix, B is $s \times s$ and C is $q \times q$. Of course, the equality $3r + s + q = n$ should hold. The algebra of matrices (11.60) is algebraically isomorphic to $M_r(\mathbb{C}) \oplus M_s(\mathbb{C}) \oplus M_q(\mathbb{C})$. In order to recover the subalgebra, the knowledge of the summands $M_r(\mathbb{C})$, $M_s(\mathbb{C})$ and $M_q(\mathbb{C})$ is not enough; we should know that the **multiplicity** of $M_r(\mathbb{C})$ is 3, while the multiplicity of $M_s(\mathbb{C})$ and $M_q(\mathbb{C})$ is 1.   □

Any subalgebra $\mathscr{A}$ of $B(\mathscr{H})$ induces a decomposition

$$\bigoplus_{k=1}^{K} \mathscr{H}_k^L \otimes \mathscr{H}_k^R, \qquad (11.61)$$

of $\mathscr{H}$ and elements of $\mathscr{A}$ have the form

$$\bigoplus_{k=1}^{K} A_k^L \otimes I_k^R, \text{ where } A_k^L \in B(\mathscr{H}_k^L) \qquad (11.62)$$

and I_k^R is the identity on $\mathscr{H}_k^R$. In this case $\mathscr{A}$ is isomorphic to

$$\bigoplus_{k=1}^{K} B(\mathcal{H}_k^L) \quad \text{and} \quad \sum_{k=1}^{K} \dim \mathcal{H}_k^L \times \dim \mathcal{H}_k^R = n.$$

The operators commuting with (11.62) have the form

$$\bigoplus_{k=1}^{K} I_k^L \otimes A_k^R \,, \text{ where } A_k^R \in B(\mathcal{H}_k^R) \tag{11.63}$$

and they form the commutant $\mathcal{A}'$ of $\mathcal{A}$.

In the decomposition of $\mathcal{A}$, we can group the terms with the same dimension $d = \dim \mathcal{H}_k^L$ and multiplicity $m = \dim \mathcal{H}_k^R$ together. Let $K(d,m)$ be the number of k's with multiplicity m and dimension d. An equivalent decomposition is

$$\mathcal{H} = \bigoplus_{(d,m)} \mathcal{H}_{(d,m)} \quad \text{and} \quad \mathcal{H}_{(d,m)} = \mathcal{H}_d \otimes \mathcal{H}_m \otimes \mathcal{H}_{m,d}, \tag{11.64}$$

where

$$\dim \mathcal{H}_d = d, \quad \dim \mathcal{H}_m = m, \quad \dim \mathcal{H}_{m,d} = K(d,m).$$

Elements of $\mathcal{A}$ in this representation have the form

$$A = \bigoplus_{(d,m)} \sum_{i=1}^{K(d,m)} A(d,m,i) \otimes I_m \otimes E_{i,i}^{m,d}, \qquad A(d,m,i) \in B(\mathcal{H}_d), \tag{11.65}$$

where $E_{i,j}^{m,d}$ is the set of matrix units in $\mathcal{H}_{m,d}$. If we denote by $P_{(d,m)}$ the projection onto $\mathcal{H}_{(d,m)}$, then $P_{(d,m)}$ is in the center of $\mathcal{A}$ and

$$P_{(d,m)}A = \sum_{i=1}^{K(d,m)} A(d,m,i) \otimes I_m \otimes E_{i,i}^{m,d}.$$

The next theorem describes the structure of unitaries such that the corresponding conjugation leaves a subalgebra globally invariant.

Theorem 11.26. *Assume that $\mathcal{A} \subset B(\mathcal{H})$ is a subalgebra and $U \in B(\mathcal{H})$ is a unitary such that $UAU^* \in \mathcal{A}$ for every $A \in \mathcal{A}$. Then it follows that in terms of the decomposition (11.65) U commutes with all projections $P_{(d,m)}$ and*

$$P_{(d,m)}U = \sum_{i=1}^{K(d,m)} U_i^L \otimes U_i^R \otimes E_{\sigma(i),i}^{m,d},$$

where $U_i^L \in B(\mathcal{H}_d)$ and $U_i^R \in B(\mathcal{H}_m)$ are unitaries and σ is a permutation of the set $\{1,2,\ldots,K(d,m)\}$.

If instead of a single unitary, we have a continuous one-parameter family U_t of unitaries such that $U_t \mathcal{A} U_t^* \subset \mathcal{A}$, then the permutation σ must be a continuous function of $t \in \mathbb{R}$. The only possibility is $\sigma=$identity. Therefore, we do not need the decomposition (11.64), the simpler (11.61) is good enough to see the structure

$$U_t = \bigoplus_{k=1}^{K} (U_k^L)_t \otimes (U_k^R)_t = \left(\bigoplus_{k=1}^{K} (U_k^L)_t \otimes I_k^R \right)\left(\bigoplus_{k=1}^{K} I_k^L \otimes (U_k^R)_t \right) \qquad (11.66)$$

where $(U_k^L)_t$ and $(U_k^R)_t$ are one-parameter families of unitaries. Note that the first factor of (11.66) is in $\mathscr{A}'$, while the second one is in $\mathscr{A}$.

Theorem 11.27. *Let $\mathscr{A} \subset B(\mathscr{H})$ be a subalgebra with decomposition (11.62) and let ρ be a density matrix. If $\rho^{it}\mathscr{A}\rho^{-it} \subset \mathscr{A}$ for every $t \in \mathbb{R}$, then ρ has the form*

$$\rho = \bigoplus_{k=1}^{K} A_k^L \otimes A_k^R$$

or

$$\rho = \rho_0 D,$$

where $\rho_0 \in \mathscr{A}$ is the reduced density of ρ and $D \in \mathscr{A}'$.

Assume that $\mathscr{M}$ is a matrix algebra. The center $\{A_0 \in \mathscr{M} : A_0 A = A A_0$ for every $A \in \mathscr{M}\}$ is an algebra as well. Suppose that it has the minimal projections $p_1, p_2, \ldots, p_m$. Then $p_i \mathscr{M}$ is isomorphic to a full matrix algebra $M_{m_i}(\mathbb{C})$ and $\mathscr{A}$ is isomorphic to the direct sum

$$M_{k_1}(\mathbb{C}) \oplus M_{k_2}(\mathbb{C}) \oplus \ldots \oplus M_{k_m}(\mathbb{C}).$$

Up to an isomorphism the algebra $\mathscr{M}$ is determined by the vector $\mathbf{k} := (k_1, k_2, \ldots, k_m)^T$. Let $\mathscr{N}$ be another matrix algebra with minimal central projections $q_1, q_2, \ldots, q_n$ and let $q_j \mathscr{N}$ be isomorphic to a full matrix algebra $M_{n_j}(\mathbb{C})$. Then $\mathscr{N}$ is determined by the vector $\ell := (l_1, l_2, \ldots, l_n)^T$.

Assume now that $\mathscr{N} \subset \mathscr{M}$. Then $p_i q_j \mathscr{N}$ is a subalgebra of $p_i \mathscr{M}$ if $p_i q_j \neq 0$. The (i,j) element of the matrix $\Lambda_{\mathscr{N}}^{\mathscr{M}}$ is 0 if $p_i q_j = 0$ and the multiplicity of $p_i q_j \mathscr{N}$ in $p_i \mathscr{M}$ otherwise. The matrix $\Lambda_{\mathscr{N}}^{\mathscr{M}}$ is called the **inclusion matrix** of the relation $\mathscr{N} \subset \mathscr{M}$. The matrix $\Lambda_{\mathscr{N}}^{\mathscr{M}}$ is an $m \times n$ matrix and has the property

$$\Lambda_{\mathscr{N}}^{\mathscr{M}} \ell = \mathbf{k}.$$

If $\mathscr{N}$ and the inclusion matrix $\Lambda_{\mathscr{N}}^{\mathscr{M}}$ are given, then the algebra $\mathscr{M}$ is determined, but the inclusion $\mathscr{N} \subset \mathscr{M}$ can be reconstructed only up to a unitary transformation.

In Example 11.29, the inclusion matrix is $[3,1,1]$, $\ell = (r,s,q)^T$ and $\mathbf{k} = (3r+s+q)$.

11.9 Conjugate Convex Function

Let V be a finite-dimensional vector space with dual V^*. Assume that the duality is given by a bilinear pairing $\langle \cdot, \cdot \rangle$. For a convex function $F : V \to \mathbb{R} \cup \{+\infty\}$ the **conjugate convex function** $F^* : V^* \to \mathbb{R} \cup \{+\infty\}$ is given by the formula

$$F^*(v^*) = \sup\{\langle v, v^* \rangle - F(v) : v \in V\}.$$

F^* is sometimes called the **Legendre transform** of F.

Theorem 11.28. *If $F : V \to \mathbb{R} \cup \{+\infty\}$ is a lower semi-continuous convex function, then $F^{**} = F$.*

Example 11.30. Fix a density matrix $\rho = e^H$ and consider the functional F

$$F(X) = \begin{cases} \mathrm{Tr}\, X (\log X - \log \rho) & \text{if } X \geq 0 \text{ and } \mathrm{Tr}\, X = 1 \\ +\infty & \text{otherwise.} \end{cases}$$

defined on self-adjoint matrices. F is essentially the relative entropy with respect to ρ.

The duality is $\langle X, B \rangle = \mathrm{Tr}\, XB$ if X and B are self-adjoint matrices.

Let us show that the functional $B \mapsto \log \mathrm{Tr}\, e^{H+B}$ is the Legendre transform or the conjugate function of F:

$$\log \mathrm{Tr}\, e^{B+H} = \max\{\mathrm{Tr}\, XB - S(X\|\rho) : X \text{ is positive}, \mathrm{Tr}\, X = 1\}. \qquad (11.67)$$

On the other hand, if X is positive invertible with $\mathrm{Tr}\, X = 1$, then

$$S(X\|\rho) = \max\{\mathrm{Tr}\, XB - \log \mathrm{Tr}\, e^{H+B} : B \text{ is self-adjoint}\}. \qquad (11.68)$$

Introduce the notation

$$f(X) = \mathrm{Tr}\, XB - S(X\|\rho)$$

for a density matrix X. When $P_1, \ldots, P_n$ are projections of rank one with $\sum_{i=1}^n P_i = I$, we can write

$$f\left(\sum_{i=1}^n \lambda_i P_i\right) = \sum_{i=1}^n (\lambda_i \mathrm{Tr}\, P_i B + \lambda_i \mathrm{Tr}\, P_i \log \rho - \lambda_i \log \lambda_i),$$

where $\lambda_i \geq 0$, $\sum_{i=1}^n \lambda_i = 1$. Since

$$\frac{\partial}{\partial \lambda_i} f\left(\sum_{i=1}^n \lambda_i P_i\right)\bigg|_{\lambda_i = 0} = +\infty,$$

we can see that $f(X)$ attains its maximum at a positive matrix X_0, $\mathrm{Tr}\, X_0 = 1$. Then for any self-adjoint Z, $\mathrm{Tr}\, Z = 0$, we have

$$0 = \frac{d}{dt} f(X_0 + tZ)\Big|_{t=0} = \mathrm{Tr}\, Z(B + \log \rho - \log X_0),$$

so that $B + H - \log X_0 = cI$ with $c \in \mathbb{R}$. Therefore $X_0 = e^{B+H}/\mathrm{Tr}\, e^{B+H}$ and $f(X_0) = \log \mathrm{Tr}\, e^{B+H}$ by simple computation.

Let us next prove (11.68). It follows from (11.67) that the functional $B \mapsto \log \mathrm{Tr}\, e^{H+B}$ defined on the self-adjoint matrices is convex. Let $B_0 = \log X - H$ and

$$g(B) = \mathrm{Tr}\, XB - \log \mathrm{Tr}\, e^{H+B}$$

which is concave on the self-adjoint matrices. Then for any self-adjoint S we have

$$\frac{d}{dt} g(B_0 + tS)\Big|_{t=0} = 0,$$

because $\mathrm{Tr}\, X = 1$ and

$$\frac{d}{dt} \mathrm{Tr}\, e^{\log X + tS}\Big|_{t=0} = \mathrm{Tr}\, XS.$$

Therefore g has the maximum $g(B_0) = \mathrm{Tr}\, X(\log X - H)$, which is the relative entropy of X and ρ.   □

Example 11.31. Let ω and ρ be density matrices. By modification of (11.68) we may set

$$S_{co}(\omega\|\rho) = \max\{\mathrm{Tr}\, \omega B - \log \mathrm{Tr}\, \rho e^B : B \text{ is self-adjoint}\}. \tag{11.69}$$

It is not difficult to see that

$$S_{co}(\omega\|\rho) = \max\{S(\omega|\mathscr{C}\|\rho|\mathscr{C}) : \mathscr{C}\} \tag{11.70}$$

where $\mathscr{C}$ runs over all commutative subalgebras. It follows from the monotonicity of the relative entropy that

$$S_{co}(\omega\|\rho) \leq S(\omega\|\rho). \tag{11.71}$$

The sufficiency theorem tells that the inequality is strict if ω and ρ do not commute.   □

11.10 Some Trace Inequalities

The **Golden–Thompson inequalilty** tells us that

$$\mathrm{Tr}\, e^{A+B} \leq \mathrm{Tr}\, e^A e^B$$

holds for self-adjoint A and B.

The Golden–Thompson inequalilty can be deduced from inequality (11.71). Putting $X = e^{A+B}/\mathrm{Tr}\, e^{A+B}$ for Hermitian A and B we have

$$\log \mathrm{Tr}\, e^A e^B \geq \mathrm{Tr}\, XA - S_{co}(X, e^B) \geq \mathrm{Tr}\, XA - S(X, e^B) = \log \mathrm{Tr}\, e^{A+B},$$

which further shows that $\mathrm{Tr}\, e^{A+B} = \mathrm{Tr}\, e^A e^B$ holds if and only if $AB = BA$.

According to Araki,

$$\mathrm{Tr}\, (X^{1/2} Y X^{1/2})^{rp} \leq \mathrm{Tr}\, (X^{r/2} Y^r X^{r/2})^p \tag{11.72}$$

holds for every number $r \geq 1$, $p > 0$ and positive matrices X, Y. (11.72) is called **Araki–Lieb–Thirring inequality** [8] and it implies that the function

$$p \mapsto \mathrm{Tr}\, (e^{pB/2} e^{pA} e^{pB/2})^{1/p} \tag{11.73}$$

is increasing for $p > 0$. Its limit at $p = 0$ is $\mathrm{Tr}\, e^{A+B}$. Hence we have a strengthened variant of the Golden–Thompson inequality.

The formal generalization

$$\mathrm{Tr}\, e^{A+B+C} \leq \mathrm{Tr}\, e^A e^B e^C$$

of the Golden–Thompson inequality is false. However, if two of the three matrices commute then the inequality holds obviously. A nontrivial extension of the Golden–Thompson inequality to three operators is due to Lieb [71].

Theorem 11.29. *Let A, B and C be self-adjoint matrices. Then*

$$\mathrm{Tr}\, e^{A+B+C} \leq \int_0^\infty \mathrm{Tr}\, (t + e^{-A})^{-1} e^B (t + e^{-A})^{-1} e^C \, dt$$

11.11 Notes

Theorem 11.11 and inequlity (11.26) are from the paper [24] of Carlen and Lieb.

The classical source about majorization is [75]. In the matrix setting [6] and [51] are good surveys. The latter discusses log-majorization as well. Theorem 11.16 was developed by A. Wehrl in 1974.

The operator monotonicity of the function (11.38) is discussed in [96, 111]. Theorem 11.18 was developed in [10]. Operator means have been extended to more than two operators in [100].

11.12 Exercises

1. Show that
$$\|x - y\|^2 + \|x + y\|^2 = 2\|x\|^2 + 2\|y\|^2 \tag{11.74}$$
   for the norm in a Hilbert space. (This called "parallelogram law.")

2. Give an example of $A \in M_n(\mathbb{C})$ such that the spectrum of A is in $\mathbb{R}^+$ and A is not positive.

3. Let $A \in M_n(\mathbb{C})$. Show that A is positive if and only if X^*AX is positive for every $X \in M_n(\mathbb{C})$.

4. Let $A \in M_n(\mathbb{C})$. Show that A is positive if and only if $\operatorname{Tr} XA$ is positive for every positive $X \in M_n(\mathbb{C})$.

5. Let $\|A\| \le 1$. Show that there are unitaries U and V such that

$$A = \frac{1}{2}(U + V).$$

(Hint: Use Example 11.2.)

6. Let $V : \mathbb{C}^n \to \mathbb{C}^n \otimes \mathbb{C}^n$ be defined as $Ve_i = e_i \otimes e_i$. Show that

$$V^*(A \otimes B)V = A \circ B \qquad (11.75)$$

for $A, B \in M_n(\mathbb{C})$. Conclude the **Schur theorem**.

7. Let $A \in M_n(\mathbb{C})$ be positive and let X be an $n \times n$ positive block-matrix (with $k \times k$ entries). Show that the block-matrix

$$Y_{ij} = A_{ij}X_{ij} \qquad (1 \le i, j \le n)$$

is positive. (Hint: Use Theorem 11.8.)

8. Let $\alpha : \mathbb{C}^n \to M_m(\mathbb{C})$ be a positive mapping. Show that α is completely positive.

9. Let $\alpha : M_n(\mathbb{C}) \to M_m(\mathbb{C})$ be a completely positive mapping. Show that its adjoint $\alpha^* : M_m(\mathbb{C}) \to M_n(\mathbb{C})$ is completely positive.

10. Give a proof for the strong subadditivity of the von Neumann entropy by differentiating inequality (11.26) at $r = 1$.

11. Let $\alpha : M_n(\mathbb{C}) \to M_n(\mathbb{C})$ be a positive unital mapping and let $0 < t < 1$. Show that for every positive matrix $A \in M_n(\mathbb{C})$, the inequality

$$\alpha(A^t) \le \alpha(A)^t$$

holds. (Hint: $f(x) - x^t$ is operator monotone function.)

12. Use the **Schur factorization** (11.6) to show that

$$\det\left(\begin{bmatrix} A & B \\ B^* & C \end{bmatrix} \right) = \det A \times \det(C - B^*A^{-1}B)$$

if A is invertible. What is the determinant if A is not invertible?

13. Deduce the subadditivity of the von Neumann entropy differentiating the inequality in Theorem 11.11 at $p = 1$.

14. Assume that the block-matrix

$$\begin{bmatrix} A & B \\ B^* & C \end{bmatrix} \qquad (11.76)$$

is invertible. Show that A and $C - B^*A^{-1}B$ must be invertible.

15. Use the factorization (11.6) to show that the inverse of the block-matrix (11.76) is

$$\begin{bmatrix} A^{-1}+A^{-1}B(C-B^*A^{-1}B)^{-1}B^*A^{-1} & -A^{-1}B(C-B^*A^{-1}B)^{-1} \\ -(C-B^*A^{-1}B)^{-1}B^*A^{-1} & (C-B^*A^{-1}B)^{-1} \end{bmatrix}. \quad (11.77)$$

16. Show that for a self-adjoint matrix A definitions (11.8) and (11.9) give the same result.

17. Use the **Frobenius formula**

$$\frac{f(s)-f(r)}{s-r} = \frac{1}{2\pi i} \int_\Gamma \frac{f(z)}{(z-s)(z-r)}\, dz \quad (11.78)$$

   to deduce (11.12) from (11.11).

18. Show that $f(x)=x^2$ is not operator monotone on any interval.

19. Deduce the inequality

$$\sqrt{xy} \le \frac{x-y}{\log x - \log y} \quad (11.79)$$

   (between the geometric and logarithmic means) from the operator monotonicity of the function $\log t$. (Hint: Apply Theorem 11.17.)

20. Use Theorem 11.17 and the formula

$$\mathrm{Det}\left(\left[\frac{1}{a_i+b_j}\right]_{i,j=1}^n\right) = \prod_{1\le i<j\le n}(a_i-a_j) \prod_{1\le i<j\le n}(b_i-b_j) \prod_{1\le i,j\le n}(a_i+b_j)^{-1} \quad (11.80)$$

   to show that the square root is operator monotone.

21. Let P and Q be projections. Show that

$$P\#Q = \lim_{n\to\infty}(PQP)^n.$$

   Is this a projection?

22. Show that $f(x)=x^r$ is not operator monotone on $\mathbb{R}^+$ when $r>1$. A possibility is to choose the real positive parameters b_1 and b_2 such that for the matrices

$$A := \begin{bmatrix} 1 & 1 \\ 1 & 1 \end{bmatrix} \quad \text{and} \quad B := \begin{bmatrix} b_1 & 0 \\ 0 & b_2 \end{bmatrix}$$

   $0 \le A \le B$ holds but $A^r \le B^r$ does not.

23. Let A and B be self-adjoint matrices and P be a projection. Give an elementary proof of the inequality

$$(PAP+P^\perp BP^\perp)^2 \le PA^2P+P^\perp B^2P^\perp,$$

   where $P^\perp$ stands for $I-P$.

24. Let $A \geq 0$ and P be a projection. Show that

$$A \leq 2(PAP + P^\perp A P^\perp),$$

where $P^\perp = I - P$.

25. Let $A \geq 0$ and P be a projection. Representing A and P as

$$A = \begin{bmatrix} A_{11} & A_{12} \\ A_{21} & A_{22} \end{bmatrix} \quad \text{and} \quad P = \begin{bmatrix} I & 0 \\ 0 & 0 \end{bmatrix}$$

show that $A \leq 2PAP + 2P^\perp A P^\perp$, where $P^\perp = I - P$.

26. Deduce (11.52) for the square root function from the properties of the geometric mean.

27. Use (11.12) to show that

$$\frac{\partial}{\partial t} e^{A+tB} \bigg|_{t=0} = \int_0^1 e^{uA} B e^{(1-u)B} \, du$$

for matrices A and B.

28. Let $\alpha : M_n(\mathbb{C}) \to M_k(\mathbb{C})$ be linear mapping given by

$$\alpha(A) = \mathrm{Tr}_2 X (I \otimes A),$$

where Tr_2 denotes the partial trace over the second factor and $X \in M_k(\mathbb{C}) \otimes M_n(\mathbb{C})$ is a fixed positive matrix. Show that α is positive. Give an example such that α is not completely positive. (Hint: Write the transpose mapping in this form.)

29. Let $A \in M_n(\mathbb{C})$ be a positive matrix and define $\mathscr{E} : M_n(\mathbb{C}) \to M_n(\mathbb{C})$ as $\mathscr{E}(D) = A \circ D$, the Hadamard product by A. Show that $\mathscr{E}$ is completely positive.

30. Let C be a convex set in a Banach space. For a smooth functional $\Psi : C \to \mathbb{R}$,

$$D_\Psi(x,y) := \Psi(x) - \Psi(y) - \lim_{t \to +0} t^{-1} \Big(\Psi(y + t(x-y)) - \Psi(y) \Big)$$

is called the **Bregman divergence** of $x, y \in C$. Let C be the set of density matrices and let $\Psi(\rho) = \mathrm{Tr}\rho \log \rho$. Show that in this case the Bregman divergence is the quantum relative entropy.

31. Show that for density matrices D and $\rho = e^H$,

$$S(D\|\rho) = \sup\{\mathrm{Tr} DB - \log \mathrm{Tr} e^{H+B} : B \text{ is self-adjoint}\}. \tag{11.81}$$

holds.

Bibliography

1. L. ACCARDI AND A. FRIGERIO, Markovian cocycles, Proc. Roy. Irish Acad., **83A**(1983), 251–263.
2. J. ACZÉL AND Z. DARÓCZY, *On measures of information and their characterizations*, Academic Press, New York, San Francisco, London, 1975.
3. P. M. ALBERTI AND A. UHLMANN, *Stochasticity and partial order. Doubly stochastic maps and unitary mixing*, VEB Deutscher Verlag Wiss., Berlin, 1981.
4. R. ALICKI AND M. FANNES, Continuity of the quantum conditional information, J. Phys A: Math. Gen. **34**(2004), L55–L57.
5. S. AMARI AND H. NAGAOKA, *Methods of information geometry*, Transl. Math. Monographs **191**, AMS, 2000.
6. T. ANDO, Majorization and inequalities in matrix theory, Linear Algebra Appl. **118**(1989), 163–248.
7. H. ARAKI, Relative entropy for states of von Neumann algebras, Publ. RIMS Kyoto Univ. **11**(1976), 809–833.
8. H. ARAKI, On an inequality of Lieb and Thirring, Lett. Math. Phys. **19**(1990), 167–170.
9. H. ARAKI AND H. MORIYA, Equilibrium statistical mechanics of fermion lattice systems, Rev. Math.Phys, **15**(2003), 93–198.
10. K. M. R. Audenaert, J. Calsamiglia, Ll. Masanes, R. Munoz-Tapia, A. Acin, E. Bagan, F. Verstraete, The quantum Chernoff bound, quant-ph/0610027, 2006.
11. O. E. BARNDORFF-NIELSEN AND P. E. JUPP, Yokes and symplectic structures, J. Stat. Planning and Inference, **63**(1997), 133–146.
12. H. BARNUM, E. KNILL AND M. A. NIELSEN, On quantum fidelities and channel capacities, IEEE Trans.Info.Theor. **46**(2000), 1317–1329.
13. V. P. BELAVKIN AND P. STASZEWSKI, C*-algebraic generalization of relative entropy and entropy, Ann. Inst. Henri Poincaré, Sec. A **37**(1982), 51–58.
14. C. H. BENNETT, G. BRASSARD, C. CREPEAU, R. JOZSA, A. PERES AND W. WOOTTERS, Teleporting an unknown quantum state via dual classical and EPR channels, Physical Review Letters, **70**(1993), 1895–1899.
15. I. BJELAKOVIĆ, *Limit theorems for quantum entropies*, Ph.D. Dissertation, Berlin, 2004.
16. I. BJELAKOVIĆ AND A. SZKOLA, The data compression theorem for ergodic quantum information sources, Quantum. Inf. Process. **4**(2005), 49–63.
17. I. BJELAKOVIĆ AND R. SIEGMUND-SCHULTZE, An ergodic theorem for the quantum relative entropy, Commun. Math. Phys. **247**(2004), 697–712.
18. I. BJELAKOVIĆ, T. KRÜGER, R. SIEGMUND-SCHULTZE AND A. SZKOŁA, The Shannon-McMillan theorem for ergodic quantum lattice systems. Invent. Math. **155**(2004), 203–222.
19. J. BLANK, P. EXNER AND M. HAVLIČEK, *Hilbert space operators in quantum physics*, American Institute of Physics, 1994.
20. R. BHATIA, *Matrix analysis*, Springer-Verlag, New York, 1996.

21. O. BRATTELI AND D. W. ROBINSON, *Operator Algebras and Quantum Statistical Mechanics II*, Springer-Verlag, New York-Heidelberg-Berlin, 1981.

22. S. L. BRAUNSTEIN AND C. M. CAVES, Statistical distance and the geometry of quantum states, Phys. Rev. Lett. **72**(1994), 3439–3443.

23. L. L. CAMPBELL, A coding theorem and Rényi's entropy, Information and Control, **8**(1965), 523–429.

24. E. A. CARLEN AND E. H. LIEB, A Minkowski type trace inequality and strong subadditivity of quantum entropy, Amer. Math. Soc. Transl. **189**(1999), 59–69.

25. M. D. CHOI, Completely positive mappings on complex matrices, Linear Algebra Appl. **10**(1977), 285–290.

26. T. M. COVER AND J. A. THOMAS, *Elements of information theory*, Wiley, 1991.

27. I. CSISZÁR, Information type measure of difference of probability distributions and indirect observations, Studia Sci. Math. Hungar. **2**(1967), 299–318.

28. I. CSISZÁR, I-divergence geometry of probability distributions and minimization problems, Ann. Prob. **3**(1975), 146–158.

29. I. CSISZÁR AND J. KÖRNER, *Information theory. Coding theorems for discrete memoryless systems*, Akadémiai Kiadó, Budapest, 1981.

30. I. CSISZÁR AND P. SHIELDS, Information theory and statistics: A tutorial, Foundations and Trends in Communications and Information Theory, **1**(2004), 417–528.

31. I. CSISZÁR, F. HIAI AND D. PETZ, A limit relation for quantum entropy and channel capacity per unit cost, arXiv:0704.0046, 2007.

32. M. CHRISTANDL AND A. WINTER,"Squashed entanglement" - An additive entanglement measure, J. Math. Phys. **45**(2004), 829–840.

33. Z. DARÓCZY, Generalized information functions, Information and Control, **16**(1970), 36–51.

34. L. DIÓSI, T. FELDMANN AND R. KOSLOFF, On the exact identity between thermodynamic and informatic entropies in a unitary model of friction Internat. J. Quantum Information **4**(2006), 99–104.

35. J. L. DODD AND M. A. NIELSEN, A simple operational interpretation of fidelity, arXiv e-print quant-ph/0111053.

36. M. J. DONALD, M. HORODECKI AND O. RUDOLPH, The uniqueness theorem for entanglement measures, J. Math. Phys. **43**(2002), 4252–4272.

37. S. EGUCHI, Second order efficiency of minimum contrast estimation in a curved exponential family, Ann. Statist. **11**(1983), 793–803.

38. M. FANNES, A continuity property of the entropy density for spin lattice systems, Commun. Math. Phys. **31**(1973), 291–294.

39. M. FANNES, J. T. LEWIS AND A. VERBEURE, Symmetric states of composite systems, Lett. Math. Phys. **15**(1988), 255–260.

40. E. FICK AND G. SAUERMANN, *The quantum statistics of dynamic processes*, Springer, Berlin, Heidelberg, 1990.

41. A. FUJIWARA AND T. HASHIZUMÉ, Additivity of the capacity of depolarizing channels, Phys. Lett. A **299**(2002), 469–475.

42. R. GILL AND S. MASSAR, State estimation for large ensembles, Phys. Rev. A., **61**(2000), 042312.

43. F. HANSEN AND G. K. PEDERSEN, Jensen's inequality for operator and Löwner's theorem, Math. Anal. **258**(1982), 229–241.

44. R. V. L. HARTLEY, *Transmission of Information*, Bell System Technical Journal, **7**(1928), 535–567.

45. M. HAYASHI, *Quantum information. An introduction*, Springer, 2006.

46. M. HAYASHI AND K. MATSUMOTO, Asymptotic performance of optimal state estimation in quantum two level systems, arXiv:quant-ph/0411073.

47. P. HAYDEN, R. JOZSA, D. PETZ AND A. WINTER, Structure of states which satisfy strong subadditivity of quantum entropy with equality, Commun. Math. Phys. **246**(2004), 359–374.

48. J. HAVRDA AND F. CHARVÁT, Quantification methods of classification processes. Concept of structural α-entropy, Kybernetika (Prague), **3**(1967), 30–35.

49. P. HAUSLADEN, R. JOZSA, B. SCHUMACHER, M. WESTMORELAND AND W. WOOTERS, Classical information capacity of a quantum channel, Phys. Rev. A **54**(1996), 1869–1876.

50. C. W. HELSTROM *Quantum detection and estimation theory.* Academic Press, New York, 1976.

51. F. HIAI, Log-majorizations and norm inequalities for exponential operators, in *Linear operators* (Warsaw, 1994), 119–181, Banach Center Publ., **38**, Polish Acad. Sci., Warsaw, 1997.

52. F. HIAI, M. OHYA AND M. TSUKADA, Sufficiency, KMS condition a relative entropy in von Neumann algebras, Pacific J. Math. **96**(1981), 99–109.

53. F. HIAI AND D. PETZ, The proper formula for relative entropy and its asymptotics in quantum probability, Commun. Math. Phys. **143**(1991), 99–114.

54. H. F. HOFMANN AND S. TAKEUCHI, Violation of local uncertainty relations as a signature of entanglement, Phys. Rev. A **68**(2003), 032103.

55. A. S. HOLEVO, Problems in the mathematical theory of quantum communication channels, Rep. Math. Phys. **12**(1977), 273–278.

56. A. S. HOLEVO, *Probabilistic and statistical aspects of quantum theory*, North-Holland, Amsterdam, 1982.

57. A. S. HOLEVO, The capacity of quantum channels with general signal states, IEEE Trans. Inf. Theory, **44**(1998), 269–273.

58. A. S. HOLEVO, Reliability function of general classical-quantum channel, IEEE Trans. Inf. Theory, **46**(2000), 2256–2261.

59. A. S. HOLEVO, *Statistical structure of quantum theory*, Lecture Notes in Phys. 67, Springer, Heidelberg, 2001.

60. M. HORODECKI, P.W. SHOR AND M. B. RUSKAI, Entanglement breaking channels, Rev. Math. Phys. **15**(2003), 629–641.

61. B. IBINSON, N. LINDEN AND A. WINTER, Robustness of quantum Markov chains arXiv:quant-ph/0611057, 2006.

62. A. JENČOVA, Generalized relative entropies as contrast functionals on density matrices, Internat. J. Theoret. Phys. **43**(2004), 1635–1649.

63. A. JENČOVA AND D. PETZ, Sufficiency in quantum statistical inference, Commun. Math. Phys. **263**(2006), 259–276.

64. A. JENČOVA AND D. PETZ, Sufficiency in quantum statistical inference: A survey with examples, J. Infinite Dimensional Analysis and Quantum Probability, **9**(2006), 331–352.

65. R. JOZSA, M. HORODECKI, P. HORODECKI AND R. HORODECKI, Universal quantum information compression, Phys. Rev. Lett. **81**(1988), 1714–1717.

66. A. KALTCHENKO AND E. YANG, Universal compression of ergodic quantum sources, Quantum Inf. Comput. **3**(2003), 359–375.

67. K. KRAUS, Complementarity and uncertainty relations, Phys. Rev D. **35**(1987), 3070–3075.

68. K. KRAUS, *States, effects and operations*, Springer, 1983.

69. F. KUBO AND T. ANDO, Means of positive linear operators, Math. Ann. **246**(1980), 205–224.

70. A. LESNIEWSKI AND M. B. RUSKAI, Monotone Riemannian metrics and relative entropy on noncommutative probability spaces, J. Math. Phys. **40**(1999), 5702–5724.

71. E. H. LIEB, Convex trace functions and the Wigner-Yanase-Dyson conjecture, Advances in Math. **11**(1973), 267–288.

72. E. H. LIEB AND M. B. RUSKAI, Proof of the strong subadditivity of quantum mechanical entropy, J. Math, Phys. **14**(1973), 1938–1941.

73. G. LINDBLAD, Completely positive maps and entropy inequalities, Commun. Math. Phys. **40**(1975), 147–151.

74. T. LINDVAL, *Lectures on the coupling method*, John Wiley, 1992.

75. A. W. MARSHALL AND I. OLKIN, *Inequalities: Theory of majorization and its applications*, Academic Press, New York, 1979.

76. M. MOSONYI AND D. PETZ, Structure of sufficient quantum coarse-grainings, Lett. Math. Phys. **68**(2004), 19–30.

77. J. VON NEUMANN, Thermodynamik quantummechanischer Gesamheiten, Gött. Nach. **1**(1927), 273–291.

78. J. VON NEUMANN, *Mathematische Grundlagen der Quantenmechanik*, Springer, Berlin, 1932. English translation: Mathematical foundations of quantum mechanics, Dover, New York, 1954.

79. M. A. NIELSEN AND J. KEMPE, Separable states are more disordered globally than locally, Phys. Rev. Lett., **86**(2001), 5184–5187.

80. M. A. NIELSEN AND D. PETZ, A simple proof of the strong subadditivity, Quantum Inf. Comp., **5**(2005), 507–513.

81. M. Nussbaum and A. Szkola, A lower bound of Chernoff type for symmetric quantum hypothesis testing, arXiv:quant-ph/0607216.

82. T. OGAWA AND H. NAGAOKA, Strong converse and Stein's lemma in quantum hypothesis testing, IEEE Tans. Inform. Theory **46**(2000), 2428–2433.

83. M. OHYA AND D. PETZ, *Quantum Entropy and Its Use*, Springer, 1993.

84. M. OHYA, D. PETZ, N. WATANABE, On capacities of quantum channels, Prob. Math. Stat. **17**(1997), 179–196.

85. K. R. PARTHASARATHY, On estimating the state of a finite level quantum system, Inf. Dimens. Anal. Quantum Probab. Relat. Top. **7**(2004), 607–617.

86. A. PERES, *Quantum theory: Concepts and methods*, Kluwer Academic, Dordrecht/ Boston, 1993.

87. D. PETZ, Quasi-entropies for states of a von Neumann algebra, Publ. RIMS Kyoto Univ. **23**(1985), 787–800.

88. D. PETZ, Quasi-entropies for finite quantum systems, Rep. Math. Phys. **21**(1986), 57–65.

89. D. PETZ, Sufficient subalgebras and the relative entropy of states of a von Neumann algebra, Commun. Math. Phys. **105**(1986), 123–131.

90. D. PETZ, Conditional expectation in quantum probability, in *Quantum Probability and Applications*, Lecture Notes in Math., **1303**(1988), 251–260.

91. D. PETZ, Sufficiency of channels over von Neumann algebras: Quart. J. Math. Oxford, **39**(1988), 907–1008.

92. D. PETZ, *Algebra of the canonical commutation relation*, Leuven University Press, 1990.

93. D. PETZ, Geometry of Canonical Correlation on the State Space of a Quantum System. J. Math. Phys. **35**(1994), 780–795.

94. D. PETZ, Monotone metrics on matrix spaces. Linear Algebra Appl. **244**(1996), 81–96.

95. D. PETZ AND CS. SUDÁR, Geometries of quantum states, J. Math. Phys. **37**(1996), 2662–2673.

96. H. HASEGAWA AND D. PETZ, Non-commutative extension of information geometry II, in *Quantum Communication, Computing, and Measurement*, ed. O. Hirota et al, Plenum, 1997.

97. D. PETZ, Entropy, von Neumann and the von Neumann entropy, in *John von Neumann and the Foundations of Quantum Physics*, eds. M. Rédei and M. Stöltzner, Kluwer, 2001.

98. D. PETZ, Covariance and Fisher information in quantum mechanics, J. Phys. A: Math. Gen. **35**(2003), 79–91.

99. D. PETZ, Complementarity in quantum systems, Rep. Math. Phys. **59**(2007), 209–224.

100. D. PETZ AND R. TEMESI, Means of positive numbers and matrices, SIAM Journal on Matrix Analysis and Applications, **27**(2006), 712–720.

101. M. B. PLENIO, S. VIRMANI AND P. PAPADOPOULOS, Operator monotones, the reduction criterion and relative entropy, J. Physics A **33**(2000), L193–197.

102. J. REHÁČEK, B. ENGLERT AND D. KASZLIKOWSKI, Minimal qubit tomography, Physical Review A **70**(2004), 052321.

103. J. ŘEHÁČEK AND Z. HRADIL, Quantification of entanglement by means of convergent iterations, Phys. Rev. Lett. **90**(2003), 127904.

104. A. RÉNYI, On measures of entropy and information, in *Proceedings of the 4th Berkeley conference on mathematical statistics and probability*, ed. J. Neyman, pp. 547–561, University of California Press, Berkeley, 1961.

105. E. SCHRÖDINGER, Probability relations between separated systems, Proc. Cambridge Philos. Soc. **31**(1936), 446–452.

106. B. SCHUMACHER, Quantum coding, Phys. Rev. A **51**(1995), 2738–2747.

107. J. SCHWINGER, Unitary operator basis, Proc. Nat. Acad. Sci. USA **46**(1960), 570–579.

108. C. E. SHANNON, *The mathematical theory of communication*, Bell System Technical Journal, **27**(1948), 379–423 and 623–656.

109. P. W. SHOR, Additivity of the classical capacity of entanglement-breaking quantum channels, J. Math. Phys. **43**(2002), 4334–4340.

110. P. W. SHOR, Equivalence of additivity questions in quantum information theory, Comm. Math. Phys. **246**(2004), 453–472.

111. V. E. S. SZABO, A class of matrix monotone functions, Linear Algebra Appl. **420**(2007), 79–85.

112. C. TSALLIS, Possible genralizationof Boltzmann-Gibbs statistics, J. Stat. Phys. **52**(1988), 479–487.

113. A. UHLMANN, The "transition probability" in the state space of a *-algebra, Rep. Math. Phys. **9**(1976), 273–279.

114. A. UHLMANN, Relative entropy and the Wigner-Yanase-Dyson-Lieb concavity in an interpolation theory, Comm. Math. Phys. **54**(1977), 21–32.

115. H. UMEGAKI, Conditional expectations in an operator algebra IV (entropy and information), Kodai Math. Sem. Rep. **14**(1962), 59–85.

116. V. VEDRAL, The role of relative entropy in quantum information theory, Rev. Modern Phys. **74**(2002), 197–234.

117. R. F. WERNER, All teleportation and dense coding schemes, J. Phys. A **35**(2001), 7081–7094.

118. R. F. WERNER AND A. S. HOLEVO, Counterexample to an additivity conjecture for output purity of quantum channels, **43**(2002), 4353–4357.

119. E. P. WIGNER AND M. M. YANASE, Information content of distributions, Proc. Nat. Acad. Sci. USA **49**(1963), 910–918.

120. W. K. WOOTERS AND W. H. ZUREK, A single quantum cannot be cloned, Nature, **299**(1982), 802–803.

Index

Theoretical and Mathematical Physics

**Quantum Information Theory
and Quantum Statistics**
By D. Petz

**An Introduction to Riemann
Surfaces, Algebraic Curves
and Moduli Spaces**
2nd enlarged edition
By M. Schlichenmaier

**Quantum Probability and
Spectral Analysis of Graphs**
By A. Hora and N. Obata

From Nucleons to Nucleus
Concepts of Microscopic Nuclear Theory
By J. Suhonen

**Concepts and Results in Chaotic Dynamics:
A Short Course**
By P. Collet and J.-P. Eckmann

The Theory of Quark and Gluon Interactions
4th Edition
By F. J. Ynduráin
